D1001379

KENNETH F. CHERRY

ASBESTOS

Engineering Management and Control

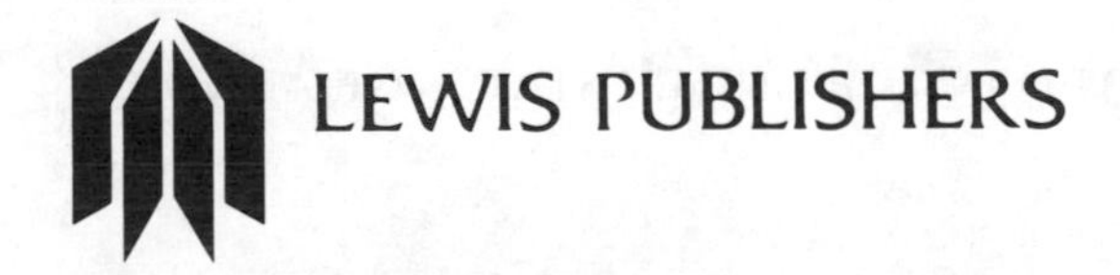

Library of Congress Cataloging-in-Publication Data

Cherry, Kenneth F.
Asbestos Engineering, Management, and Control

Bibliography: p.
Includes index.
1. Asbestos in building—Safety Measures.
2. Asbestos—Safety measures. I. Title.
TA455.A6C48 1987 690'.22 87-29836
ISBN 0-87371-127-0

LEWIS PUBLISHERS, INC.
121 South Main Street, Chelsea, Michigan 48118

PRINTED IN THE UNITED STATES OF AMERICA

This book is dedicated to those who understand the futility of approaching a goal asymptotically.

Preface

Asbestos has become a major issue, with political, economic, and technical topics frequently being addressed. Risk factors stemming from the disease-causing properties have resulted in continuing large expenditures in both the public and private sectors. Questions relating to school safety, property values, and even the continued use of vehicles with asbestos in the brake linings are all related to the presence of what was once called "nature's wonder fiber"—asbestos.

In an attempt to protect us from the danger, governmental regulatory agencies have promulgated rules, regulations, and guidelines which in turn have resulted, predictably, in confusion for regulated parties and opportunities for charlatans claiming to have all the answers. This text is intended as an instructional tool and field guide to assist in evaluating asbestos hazards, in complying with regulations, and in using a sound engineering approach to effectively deal with asbestos.

Three organizations are usually cited in asbestos hazard abatement: the U.S. Environmental Protection Agency (EPA), the Occupational Safety and Health Administration (OSHA), and the National Asbestos Council (NAC). This book is intended as a single source which presents the viewpoints, rules, and recommended practices of all three organizations.

The reader should find that this one source contains most of the background needed to address an asbestos problem as safely and as economically as possible. Two cautions are in order, however:

(1) Always consult with experienced certified industrial hygienists (CIH) certified by the American Board of Industrial Hygiene; avoid noncertified persons or those calling themselves "core certified" (core certified is a misleading term used by some industrial hygienists in training); at least, check the background of the expert you choose. (2) Always refer to the latest federal and state guidelines and check with the EPA, OSHA, and state agencies to ensure that recent changes will not be a cause for legal action against you.

Kenneth Cherry's educational background is in engineering, economics, and law. His credentials include certification as an Environmental Professional (CEP), diplomate status with the American Board of Industrial Hygiene (CIH), certification as a Hazardous Material Manager (CHMM), and registration as a professional engineer in several states (PE). He is licensed to practice patent law in the United States and Canada, holds the rank of professor with the National Institute for Environmental Communications, and has professional level membership in the American Society of Safety Engineers.

Mr. Cherry has published numerous books and articles and has served on the faculty of the Medical College of Ohio. He has taught at the University of Toledo, lectured nationally and internationally, and has served as the managing director of The National Center for Environmental Communications. With over 15 years of experience in all phases of environmental matters, including asbestos control, he has also worked for or consulted for the U.S. Army and Navy, IBM, Owens Corning Fiberglass, General Motors, The Veterans Administration, Ford Motor Company, Clayton Environmental Consultants, Davy McKee Inc., Union Carbide, several school systems, and scores of other clients. In addition, he is licensed as an asbestos abatement specialist and authorized by the National Asbestos Council as a trainer.

Contents

1
Current Uses of Asbestos

United States asbestos consumption has been declining for years because of increased restrictions and lawsuits by victims of lung cancer and other diseases. Consumption in 1985 was 155,500 metric tons, down from 226,000 in 1984 and 349,000 in 1981, according to the U.S. Bureau of Mines.[1]

The chemical process industries have gradually switched to substitutes over the years. Asbestos was long ago replaced by such materials as fiberglass and polyurethane in insulation. One area where asbestos is still widely employed is that of gaskets and packings, although a variety of substitutes is now available.

Many states have adopted an information standard for postasbestos abatement of 0.01 total fibers per cubic centimeter (f/cm^3) or less as the air quality level to be attained before the asbestos removal work area can be reoccupied. This standard is based on phase-contrast microscopy, P & CAM–239 or NIOSH 7400 (see Appendix to OSHA Part 1910 Asbestos Standard).

European countries generally have their own workplace standards for asbestos exposure, but those in the European community must also comply with a directive of 0.1 mg/m^3.

West Germany has complex regulations for the control of asbestos, and a cooperative government/industry program that began in 1981 with the goals of developing substitute materials and eventually phasing out asbestos. Roughly 70% of the asbestos used in Western Europe is reportedly used for reinforcing cement in construction.[1]

Japan's asbestos consumption nose-dived from 100,000 metric tons in fiscal 1980 to 55,000 metric tons in fiscal 1985. The Ministry of International Trade and Industry attributes this to a switch to such substitutes as corrugated metal sheets for roof tiles, and fiberglass in general for housing applications.

A problem in replacing asbestos in building materials relates to fiber strength. Asbestos-cement test sheets show a strength of 30 mega pascal (MPa), compared to 27 for glass, 25 for polyacrylonitrile, 22 for polypropylene, 20 for cellulose, 17 for polyvinyl acetate (PVA), and 17 for other mineral fibers.

Finding materials with a mechanical strength comparable to that of asbestos is difficult. Some replacements include flexible graphite, carbon and graphite yarn, glass fibers, aramid, polybenzimidazole (PBI), polytetrafluoroethylene (PTFE), ceramics, mica, and various metals.

No single material matches asbestos's strength, chemical resistance, and flexibility over a wide temperature range. Equal or superior properties may be obtained, however, by combining two or three materials.

Asbestos still has well over 50% of the packing and gasket market, compared to the traditional 80–90%. Pipe flanges are the biggest application for sealing material in chemical process plants, the others being valves, pumps, compressors, and pressure vessels. Users are reluctant to change to costlier materials with which they have had no experience. Flexible graphite can withstand temperatures of more than 5000°F (compared with about 1200°F for asbestos), say suppliers, while cautioning that it is limited to about 750–850°F in the presence of oxygen and to around 1200°F in steam.

Flexible graphite, developed about 20 years ago, was little used until recently because it costs too much. Now it has around 5–10% of the packing market. One drawback is that graphite, an electrical conductor, can enhance galvanic corrosion in steel.

In the gasket area, chlorite/graphite sheet, on the market since 1983, is the most popular nonasbestos filler for spiral-wound gaskets. In compressed gaskets—conventional asbestos encapsulated in a rubber binder—aramid, glass fibers, and mica are being used as reinforcement, while such materials as talc, barium sulfate, clay, and mica are fillers.[2]

SUMMARY OF OSHA PART 1910

The following is a brief summary of OSHA asbestos rules. It is provided to assist in explaining these regulations to nonexperts. A more complete section for technical use is provided in Chapter 9 of this book.

A. Permissible Exposure Limit (PEL)

No employee is to be exposed to airborne concentrations of asbestos in excess of 0.2 f/cm^3 of air as an 8-hr time-weighted average (TWA).

B. Monitoring

1. Every employer shall perform initial monitoring on employees who are or may be exposed at or above the action level (0.1 f/cm^3).
2. The employer may rely on objective data that demonstrates that asbestos is not capable of being released in concentrations in excess of the action level in lieu of sampling.
3. Monitoring is to be repeated at least every six months for employees whose exposures may reasonably be foreseen to exceed the action level.
4. All samples taken to satisfy the monitoring requirements shall be personal samples.
5. The employer is required to notify the affected employees of the results of the monitoring within 15 days of the receipt of the results. This must be done in writing.

C. Regulated Areas

1. The employer is required to establish regulated areas wherever airborne concentrations of asbestos exceed the permissible exposure limit.

2. These regulated areas must be demarcated in any manner that minimizes the number of persons who will be exposed to asbestos.
3. Access to regulated areas is limited to authorized persons.
4. Each person entering a regulated area shall be provided and required to use a respirator.
5. Eating, drinking, smoking, the chewing of tobacco or gum, and the application of cosmetics is prohibited in the regulated area.

D. Methods of Compliance

1. The employer is required to institute engineering and work practice controls to reduce and maintain exposures to or below the PEL.
2. All hand-operated and power-operated tools which would produce or release fibers of asbestos must be provided with local exhaust ventilation.
3. Where practical, asbestos is to be handled, mixed, applied, removed, or otherwise worked in a wet state.
4. Materials containing asbestos shall not be applied by spray methods.
5. Compressed air is not to be used to remove asbestos unless it is used in conjunction with a ventilation system.

E. Compliance Program

1. Where the PEL is exceeded, the employer is required to establish and implement a written program to reduce employee exposures to or below the PEL.
2. The employer shall not use employee rotation as a means of compliance with the PEL.

F. Respiratory Protection

1. The employer is required to provide and ensure the use of respirators where required.

2. Appropriate respirators are to be provided at no cost to the employees.
3. Power air-purifying respirators (PAPRs) are to be provided if an employee chooses to use this type and it provides adequate protection.
4. Where respirators are required, a written respirator program shall be instituted.
5. Employees who use filter respirators are to be permitted to change filters whenever an increase in breathing resistance is detected.
6. Employees who wear respirators are to be allowed to leave the regulated area whenever necessary to wash their faces and respirator facepieces to prevent skin irritation.
7. Employees shall not be assigned to tasks which require the use of respirators if an examining physician determines that the employee will be unable to function normally wearing a respirator.
8. Quantitative or qualitative fit tests shall be performed on employees wearing negative-pressure respirators at the time of initial fitting and at least every six months thereafter.

C. Protective Work Clothing and Equipment

1. Where employees are exposed to asbestos above the PEL, the employer is required to provide to employees, at no cost, appropriate protective work clothing and equipment.
2. The employer shall ensure that work clothing contaminated with asbestos is removed only in the change rooms provided.
3. Employees are not permitted to remove contaminated work clothing from the change room.
4. Contaminated work clothing shall be stored in closed containers.
5. Containers of contaminated protective equipment or work clothing are to be properly labeled.
6. Clean protective clothing is to be provided at least weekly to each affected employee.

H. Hygiene Facilities

1. The employer is required to provide clean change rooms for employees who are exposed to asbestos in excess of the PEL.
2. Employees who are exposed to asbestos in excess of the PEL are required to shower at the end of the shift.
3. The employer must provide a positive-pressure filtered-air lunch room for employees whose asbestos exposure exceeds the PEL.

I. Communication of Hazards

1. Warning signs shall be provided and displayed at each regulated area.
2. Warning labels complying with 1910.1200 shall be affixed to all asbestos-containing materials.
3. Employers who manufacture asbestos products are required to develop Material Safety Data Sheets complying with 1910.1200.
4. The employer is required to institute a training program for employees who are exposed to airborne concentrations of asbestos at or above the action level.

J. Housekeeping

1. All surfaces shall be maintained as free as practicable from accumulations of asbestos.
2. Surfaces are not to be cleaned by the use of compressed air.

K. Medical Surveillance

1. The employer must institute a medical surveillance program for employees who are or may be exposed to asbestos at or above the action level.
2. Before an employee is assigned to a job exposed to asbestos, a preplacement medical exam shall be provided.
3. Periodic medical exams are to be provided annually.
4. A medical exam is to be given within 30 calendar days before or after the date of termination of employment to any employee who has been exposed at or above the action level.

L. Recordkeeping

1. The employer is required to keep an accurate record of all measurements taken to monitor employee exposure to asbestos.
2. The employer is required to establish and maintain an accurate record for each employee subject to medical surveillance.
3. These records shall be maintained for at least 30 years.

M. Application of CFR 1910

(a) Scope and Application

This section applies to all construction work as defined in 29 CFR 1910.12 (b), including but not limited to the following:

(1) demolition or salvage of structures where asbestos, tremolite, anthophyllite, or actinolite is present

(2) removal or encapsulation of materials containing asbestos, tremolite, anthophyllite, or actinolite

(3) construction, alteration, repair, maintenance, or renovation of structures, substrates, or portions thereof, that contain asbestos, tremolite, anthophyllite, or actinolite

(4) installation of products containing asbestos, tremolite, anthophyllite, or actinolite

(5) asbestos, tremolite, anthophyllite, and actinolite spill/emergency cleanup

(6) transportation, disposal, storage, or containment of asbestos, tremolite, anthophyllite, or actinolite or products containing asbestos, tremolite, anthophyllite, or actinolite on the site or location at which construction activities are performed.

(b) Definitions

"Action level" means an airborne concentration of asbestos, tremolite, anthophyllite, actinolite, or a combination of these minerals of 0.1 f/cm^3 or air calculated as an 8-hr time-weighted average.

"Asbestos" includes chrysotile, amosite, crocidolite, tremolite asbestos, anthophyllite asbestos, actinolite asbestos, and any of these minerals that have been chemically treated and/or altered.

"Fiber" means a particulate form of asbestos, tremolite, anthophyllite, or actinolite, 5 μm or longer, with a length-to-diameter ratio of at least 3:1.

"Assistant Secretary" means the Assistant Secretary of Labor for Occupational Safety and Health, U.S. Department of Labor, or designee.

"Authorized person" means any person authorized by the employer and required by work duties to be present in regulated areas.

"Competent person" means one who is capable of identifying existing asbestos, tremolite, anthophyllite, or actinolite hazards in the workplace and who has the authority to take prompt corrective measures to eliminate them, as specified in 29 CFR 1926.32 (f). The duties of the competent person include at least the following: establishing the negative-pressure enclosure, ensuring its integrity, and controlling entry to and exit from the enclosure; supervising any employee exposure monitoring required by the standard, ensuring that all employees working within such an enclosure wear the appropriate personal protective equipment, are trained in the use of appropriate methods of exposure control, and use the hygiene facilities and decontamination procedures specified in the standards; and ensuring that engineering controls in use are in proper operating condition and are functioning properly.

"Clean room" means an uncontaminated room having facilities for the storage of employees' street clothing and uncontaminated materials and equipment.

"Decontamination area" means an enclosed area adjacent and connected to the regulated area and consisting of an equipment room, shower area, and clean room, which is used for the decontamination of workers, materials, and equipment contaminated with asbestos, tremolite, anthophyllite, or actinolite.

SUBSTANCE TECHNICAL INFORMATION FOR ASBESTOS (as published by OSHA)[3]

Substance Identification

"Asbestos" is the name of a class of magnesium silicate minerals that occur in fibrous form. Minerals that are included in this group are chrysotile, crocidolite, amosite, anthophyllite asbestos, tremolite asbestos, and actinolite asbestos.

Asbestos, tremolite, anthophyllite, and actinolite are used in the manufacture of heat-resistant clothing, automotive brake and clutch linings, and a variety of building materials including floor tiles, roofing felts, ceiling tiles, asbestos-cement pipe and sheet, and fire-resistant drywall. Asbestos, tremolite, anthophyllite, and actinolite are also present in pipe and boiler insulation materials and in sprayed-on materials located on beams in crawl spaces and between walls.

The potential of an asbestos-containing product for releasing breathable fibers depends on its degree of friability. "Friable" means that the material can be crumbled with hand pressure and is therefore likely to emit fibers. The fibrous, fluffy sprayed-on materials used for fireproofing, insulation, or soundproofing are considered to be friable and they readily release airborne fibers if disturbed. Materials such as vinyl-asbestos floor tile or roofing felts are considered nonfriable and generally do not emit airborne fibers unless subjected to sanding or sawing operations. Asbestos-cement pipe or sheet can emit airborne fibers if the materials are cut or sawed or if they are broken during demolition operations.

Permissible Exposure

Exposure to airborne asbestos, tremolite, anthophyllite, and actinolite fibers may not exceed 0.2 f/cm^3 of air averaged over the 8-hr workday by OSHA standards.

[illegible] crocidolite [illegible] amosite, [illegible] asbestos, and actinolite asbestos.

[illegible]

[illegible] between walls.

The potential of an asbestos-containing product for releasing breathable fibers depends [illegible] friable [illegible] crumbled with hand pressure and [illegible] are generally [illegible] considered to be [illegible] fibers [illegible] unless subjected to sanding or sawing [illegible]

References [illegible]

[illegible]

2
Engineering Aspects

POINTS TO CONSIDER

A sometimes overlooked aspect of abatement projects is the opportunity to improve interior design. Large-scale asbestos abatement projects are common, and millions of dollars are being spent on them. Asbestos abatement consultation is a lucrative but demanding profession. Even the most knowledgeable asbestos abatement experts sometimes fail to recognize the impact of their work on the physical surroundings. The cost implications of new wall, floor, and ceiling finishes, furnishings, and systems usually go unnoticed.

The nightmare of totally gutting an old high-rise building down to the basic structural elements can be overcome by the design of a more functional, more efficient, and more pleasant interior space. Large-scale asbestos removal projects need to be approached as if the abatement activity is creating the opportunity for better designs.

Quality interior design promotes productivity of office workers, resulting in enhanced efficiency for owners and executives. Energy conservation opportunities also exist. Proof of this can be demonstrated with life-cycle costing. For the building owner who may face an asbestos risk but cannot deal with the enormous costs of removal, the life-cycle costing arguments are often quite convincing.

For example, we estimate that an asbestos maintenance program costing $2/ft^2 will extend the useful life of the asbestos insulation for five years. Also, a removal and putback cost of $20/ft^2 will eliminate the potential asbestos problem and extend the service life of the building by 20 years. To compare the two alternatives requires that each capital cost be reduced to Equivalent Annual Cost (EAC).

Assuming an interest rate of 12%, the annual cost for the repair and encapsulation is about EAC = $0.60/ft^2. For the removal and renovation costs, EAC = $2.70/ft^2.

Assume two additional parameters: (1) energy costs for the facility have been improved by improving heating, ventilating, and air conditioning (HVAC) equipment ... a savings of $0.50/ft^2 annually; and (2) improved interior space results in more efficient personnel, reducing capital by $1.75/ft^2. Then $2.70 – 0.50 – 1.75 = $0.45/ft^2.

The more expensive initial investment of $20/ft^2 for removal and renovation is cost-effective and the asbestos problem is eliminated. As a result, the owner can: (1) save the difference of $0.60 – 0.45 = $0.15/ft^2 annually; (2) eliminate the asbestos problem; and (3) have a more attractive interior space.

An asbestos abatement project is basically a construction project with many of the components used in demolition and renovation work. The contractual relationships between owners, architects, and contractors are the same as those used in new construction. The process often starts with the owner engaging the services of an engineer to determine the project requirements and preparing a set of bidding and contract documents containing drawings and specifications, issued for competitive bidding. (Chapter 6 in this book shows examples of specifications used by some large building owners.) During the course of construction, the owner will frequently hire a project manager who will ensure that the work is performed according to the requirements of the drawings and specifications.

This person may be authorized to make decisions with regard to stopping the work for additional directions to the contractor on the owner's behalf. Design professionals, particularly an experienced construction manager, are invaluable.

Additional consultation may be necessary with professionals having knowledge of asbestos in buildings. Industrial hygienists, risk assessment specialists, air sampling technicians, indoor air quality experts, environmental consultants, and engineers are among the broad range of professionals with whom contact may be required. If the project manager does not have these professionals on his team, he is not a good choice to handle an asbestos job.[4]

Selection Economics

In replacing thermal insulation, economic payback is usually calculated to choose the proper insulating value for the material selected. Each user has different requirements; however, a preliminary determination can be made based on a few assumptions. While inflation, interest rates, projected fuel costs, and several other factors can be used for formal large-scale projects, an abbreviated method is often adequate for the selection.

First, determine the actual fuel cost per million Btus. Usually a phone call to the supplier will result in a valid value. Because combustion processes are not as efficient as electric steam production, multiply the fuel Btu cost by 1.2. Although the efficiency at the end user is higher for electric heat, the cost per Btu is usually much higher than direct steam from fuel processes. The reason electricity is usually a higher cost source is because the utility companies are often only about 35% efficient in converting coal's Btus into electric power.

Once the cost per Btu is determined, the installed cost of insulation must be found. An easy method for a quick estimate is to use one of the estimating guides published every year, such as "Means Construction Cost Data" (to order a copy telephone 617-747-1270). Using this source, 1 in. thick fiberglass on 3 in. pipe costs $4.60/ft, and for 2 in. thick material, $5.10/ft, while 3 in. thick material will cost about $6.20/ft.

To find the simple payback:

Heat loss in 1000 Btus/ft/hr without insulation	HLN
Heat loss in 1000 Btus/ft/hr with insulation	HLI
Cost of insulation applied/ft	CI
Value of each 1000 Btus	CH
Hours of use/yr	T

Cost/yr without insulation	$= HLN \times CH \times T = DLN$
Cost/yr for heat loss	$= HLI \times CH \times T = DLI$
Dollars/yr saved with insulation	$= DLN - DLI = DS$
Payback	$= CI/DS = PB$

Doing this calculation for each material being considered and for each commercially available thickness will allow comparison of the cost effectiveness of the alternatives.[5] Tables 1–8 in Appendix A list a variety of substitutes for asbestos-containing materials.[6]

Ventilation

One important consideration in most removal processes is the containment of contamination. The basic containment structure used usually requires "negative air machines." There are several vendors supplying these units which are simply large vacuum cleaners with filters capable of removing asbestos fibers. The particulate filters are high-efficiency particulate air (HEPA) type, which means the filter has been tested to efficiently remove most of the asbestos fibers considered dangerous.

Two things cause special problems with commercial vacuum units: buildings under negative air, and large areas. Building negative air is a condition which exists in most commercial and industrial buildings because of the design of the HVAC systems.

In an ideal situation, fans are used to supply air from outside the building through heat exchangers and duct work to occupied areas. Then exhaust ducts and fans sized to match the flow of incoming air remove stale air. In most commercial operations a portion of the air is recirculated for energy conservation. Complicating factors, such as combustion air for boilers, local exhaust for individual processes, and others, result in extreme variations between designs. Also, most older buildings have had fan or fan motor changes, as well as other equipment changes, which the original design did not take into consideration.

When filtered air exhausters are used, the building pressure should be determined during the preliminary survey. A simple gauge or incline manometer with about 20 ft of tubing is all that is required. One end of the hose is put well into the building, the door closed, and any windows and doors likely to be closed during removal should also be closed. The other end is placed outside 4 ft or more from the opening, and the gauge is read in inches of water.

To illustrate the next procedure, assume that a negative pressure (vacuum) exists in the building of 0.3 in. of water column. Also, we will assume that the interior of the containment area is 15,000 ft^3. Designing for one air change per 30 min means exhausting 500 ft^3 per min. The air lock is about 5 ft wide, 6 ft high, and 10 ft long. Because 5 ft $\times$ 6 ft gives 30 ft^2 of area, this means that if a perfect seal for the containment is made, a velocity of 17 ft/min will be found in the air lock. As a rule of thumb, any value over about 10 ft/min is adequate.

In inspecting the fan curve of a 750 ft^3/min exhauster, we find that at 0.3 in. of water, only 150 ft^3/min is exhausted. At this point there are three choices: raise the pressure in the building by readjusting the HVAC system, obtain more exhauster units, or boost the output of the available unit. Only in rare cases is there an opportunity to change the existing HVAC system, and added exhausters are very expensive.

The engineering solution is to boost the flow through the available exhauster. A booster fan is an easy and effective solution. A simple 16 in. diameter 1/4 hp fan can exhaust approximately 600 ft^3/min at 0.3 in. of vacuum. With some duct tape, plastic, and stiff wire, an adaptor to put the fan in line with the HEPA exhauster and the system will easily pull 600 ft^3/min.

Other methods successfully employed include: (1) using window fans to add air pressure to a room or area, (2) constructing a second containment area around the primary containment and using a fan to pressurize this outer shell, and (3) constructing an exterior room in which to place the exhauster.

Spray—Poly

Several vendors now offer an alternative to plastic sheets for containment construction. These materials are usually sprayed or troweled on, and when applied have the consistency of petroleum jelly. When cured, they have a rubbery feel and can be applied over windows, walls, floors, and even over soil. High material cost is somewhat offset by labor savings; however, the material can be difficult to remove and it sometimes damages painted surfaces. In addition, the fumes from some formulations may be flammable and cause headaches.

In general, most contractors find the plastic enclosures more cost-effective for asbestos containment; however, the spray-on materials can be very effective in some special applications.[5]

NATIONAL FIRE PROTECTION ASSOCIATION CODES

NFPA 220 (Edited)

Standard on Types of Building Construction

Purpose. This standard outlines basic definitions for standard types of building construction for reference by committees operating under the procedures of the National Fire Protection Association.

Scope. This standard considers only those factors necessary to define building types. The requirements for partitions, fire separation partitions, shaft enclosures, and fire walls, other than bearing walls and partitions, are not related to the construction types and need to be specified in other standards and codes, where appropriate. It is also necessary for the user to consider the influence of location, occupancy, exterior exposure, possibility of mechanical damage to fire protection material, and other features which may impose additional requirements for safeguarding life and property, as commonly covered in building codes.

Guide to Classification of Types of Construction

The types of construction include five basic types designated by Roman numerals as Type I, Type II, Type III, Type IV, and Type V. This system of designating types of construction also includes a specific breakdown of construction type through the use of Arabic numbers. These numbers follow the Roman numeral notation when naming a type of construction; e.g., Type I–443, Type II–111, Type III–200, etc.

The Arabic numbers that follow each basic type (Type I, Type II, etc.) designate the fire resistance rating requirements for certain structural elements as follows: first Arabic number—exterior bearing walls; second Arabic number—structural frame or columns and girders, supporting loads for more than one floor; and third Arabic number—floor construction.

Definitions

Fire Resistance Rating. The time, in minutes or hours, that materials or assemblies have withstood a fire exposure, as established in accordance with the test procedures of NFPA 251.

Flame Spread Rating. Numbers or classifications obtained according to NFPA 255.

Minimum Hourly Fire Resistance Rating. That degree of fire resistance deemed necessary by the authority having jurisdiction.

Types of Construction

Type I. Type I construction is that type in which the structural members, including walls, columns, beams, floors, and roofs, are of approved noncombustible or limited-combustible materials.

Table 1. Samples of Some Fire Resistance Requirements for Type I through Type IV Construction

	Type I	Type II	Type III	Type IV
Exterior bearing walls	4	2	2	2
Interior bearing walls	4	2	1	2
Columns	4	2	1	-
Beams, girders, trusses, and arches	4	2	1	-
Floor construction	3	2	1	-
Roof construction	2	1	1	-
Exterior nonbearing walls	0	0	0	-

Type II. Type II construction is that type not qualifying as Type I construction in which the structural members including walls, columns, beams, floors, and roofs are of approved noncombustible or limited-combustible materials.

Type III. Type III construction is that type in which exterior walls and structural members which are portions of exterior walls are of approved noncombustible or limited-combustible materials; and interior structural members, including walls, columns, beams, floors, and roofs, are wholly or partly of wood with smaller dimensions than required for Type IV construction or of approved noncombustible, limited-combustible, or other approved combustible materials.

Type IV. Type IV construction is that type in which exterior and interior walls and structural members which are portions of such walls are of approved noncombustible or limited-combustible materials. Other interior structural members including columns, beams, arches, floors, and roofs are of solid or laminated wood without concealed spaces and comply with the provisions of 3–4.2 through 3–4.6.

Exception: Interior columns, arches, beams, girders, and trusses of approved materials other than wood are permitted, provided they are protected to afford a fire resistance rating of not less than 1 hr.

Table 1 lists sample fire resistance requirements for Type I through Type IV construction.

NFPA 241: Standard for Safeguarding Construction, Alteration, and Demolition Operations (Edited)

Introduction

Fires during construction, alteration, or demolition operations are an ever-present threat. The fire potential is inherently greater during these operations than in the completed structure, due to previous occupancy hazard and the presence of large quantities of combustible materials and debris, together with such ignition sources as temporary heating devices, cutting/welding/plumber's torch operations, open fires, and smoking. The threat of arson is also greater during construction and demolition operations due to the availability of combustible materials on-site and the open access.

Fires during construction, alteration, or demolition operations can be eliminated or controlled through early planning, scheduling, and implementation of fire prevention measures, fire protection systems, rapid communications, and on-site security. An overall construction or demolition firesafety program shall be developed. Essential items to be emphasized include:

- good housekeeping
- on-site security
- installation of new fire protection systems as the construction progresses
- preservation of existing systems during demolition
- the organization and training of an on-site fire brigade
- rapid communication
- consideration of special hazards resulting from previous occupancies

A firesafety program shall be included in all construction, alteration, or demolition contracts, and the right of the owner to administer and enforce this program shall be established, even though the building may be entirely under the jurisdiction of the contractor.

This standard presents measures for preventing or minimizing fire damage during construction, alteration, and demolition operations. The public fire department and other fire protection authorities shall also be consulted for guidance. The unique and dangerous situations confronting fire fighters during such operations require that a complete exchange of pertinent information be established and continued during the life of the project.

General requirements applying to construction and demolition are contained in NFPA 241 Chapters 1 through 5; specific requirements for construction activities are found in NFPA 241 Chapter 6; and those requirements specific to demolition activities are covered in Chapter 7 of that document. Alteration activities may require the use of both the demolition and construction activity requirements, as applicable.

Temporary Offices and Sheds

Temporary offices, trailers, sheds, and other facilities for the storage of tools and materials, when located within the building, on the sidewalk bridging, or within 30 ft (9.1 m) of the structure, shall be of noncombustible construction. Detachment between temporary structures, adequate temporary fire protection fixed systems, and/or portable equipment, shall be provided as required by the authority having jurisdiction.

Only safely installed approved heating devices shall be used in temporary offices and sheds. Ample clearance shall be provided around stoves, heaters, and all chimney and vent connectors to prevent ignition of adjacent combustible materials, as per NFPA 211, NFPA 54, and NFPA 31.

Temporary Enclosures

Only noncombustible panels or flame-resistant tarpaulins, or other approved materials of equivalent fire-retardant characteristics shall be used. Any other fabrics or plastic films used shall be certified to conform to NFPA 701, Standard Methods of Fire Tests for Flame Resistant Textiles and Films.

Utilities

Electrical. All construction operation electrical wiring and equipment for light, heat, or power purposes shall be in accordance with pertinent provisions of NFPA 70, National Electrical Code.

Temporary Wiring. All branch circuits shall originate in an approved power outlet or panelboard. Conductors shall be permitted within multiconductor cord or cable assemblies, or as open conductors. All conductors shall be protected by overcurrent devices at their rated amperage. Runs of open conductors shall be located where the conductors will not be subject to physical damage, and the conductors shall be fastened at intervals not exceeding 10 ft (3 m). Each branch circuit that supplies receptacles or fixed equipment shall contain a separate equipment grounding when run as open conductors.

Lighting. Temporary lights shall be equipped with guards to prevent accidental contact with the bulb, except that guards are not required when construction of the reflector is such that the bulb is deeply recessed.

Temporary lighting fixtures, such as quartz, which operate at temperatures capable of igniting ordinary combustibles, shall be securely fastened so that the possibility of their coming in contact with such materials is precluded.

Temporary lights shall be equipped with heavy-duty electric cords with connections and insulation maintained in safe condition. Temporary lights shall not be suspended by their electric cords unless cords and lights are designed for this means of suspension. Splices shall have insulation equal to that of the cable.

Removal. Temporary wiring shall be removed immediately upon completion of construction or purpose for which the wiring was installed.

Fire Protection[7]

Owner's Responsibility for Fire Protection. The owner shall designate a person to be responsible for the fire prevention program and for ensuring that it is carried out to completion of the project. This fire-prevention program manager shall have the authority to enforce the provisions of this and other applicable fire-protection standards.

Where guard service is provided, the manager shall be responsible for the guard service in accordance with NFPA 601, Standard for Guard Service in Fire Loss Prevention, and NFPA 601A, Standard for Guard Operations in Fire Loss Prevention.

The manager shall be responsible to assure that proper training has been provided for the use of protection equipment. The manager shall be responsible for the presence of adequate numbers and types of fire protection devices and appliances and for their proper maintenance.

Access for Fire Fighting. A suitable location at the site shall be designated as a command post and provided with plans, emergency information, keys, communication, and equipment, as needed. The person in charge of fire protection shall respond to the location whenever fire occurs.

Every building shall be accessible to fire department apparatus by way of access roadways with all-weather driving surface of not less than 20 ft (6.1 m) of unobstructed width, to withstand the live loads of fire apparatus and having a minimum of 13 ft 6 in. (4.1 m) of vertical clearance. Dead-end fire department access roads in excess of 150 ft (45.75 m) length shall be provided with approved provisions for the turning around of fire department apparatus.

Exception: The requirement of this section may be modified when, in the opinion of the fire department, fire fighting or rescue operations would not be impaired.

First-Aid Fire Equipment. The suitability, distribution, and maintenance of extinguishers shall be in accordance with NFPA 10, Standard for Portable Fire Extinguishers. Wherever a toolhouse, storeroom, or other shanty is located in or adjacent to the building under construction or demolition, or a room or space within that building is used for storage, dressing room, or workshop, at least one approved extinguisher shall be provided and maintained in an accessible location.

Exception: This requirement may be waived if structures do not exceed 150 ft^2 floor area.

3

Establishing an Operations and Maintenance Program for Asbestos

If asbestos-containing materials (ACM) are found in a building, a special operations and maintenance (O & M) program should be implemented as soon as possible. An O & M program is recommended for each type of ACM: surfacing material, pipe and boiler insulation, and miscellaneous materials. Although many of the procedures are the same, certain steps vary according to the type of ACM.

SUMMARY

Purpose of a special O & M program: The program is designed to (1) clean up asbestos fibers previously released, (2) prevent future release by minimizing ACM disturbance or damage, and (3) monitor the condition of ACM. The program should continue until all ACM is removed or the building is demolished.

Who should participate: The asbestos program manager, the manager of building maintenance, and the supervisor of the custodial staff are key participants in the O & M program.

Program elements: The program should alert workers and building occupants to the location of ACM, train custodial and maintenance personnel in proper cleaning and maintenance, implement initial and periodic cleaning using special methods (for surfacing materials and pipe and boiler insulation only), establish a process that assures ACM is not disturbed during building repairs and renovations, and periodically reinspect areas with ACM.

Purpose of a Special O & M Program

The discovery of ACM in buildings raises two concerns: (1) how to clean up asbestos fibers previously released, and (2) how to avoid ACM disturbance or damage. The special O & M program addresses both of these issues, with procedures tailored to each of the three types of ACM.

The asbestos program manager develops and implements the special O & M program. He or she may serve as coordinator or delegate that responsibility to the facilities manager or other appropriate employee. The manager of building maintenance and the custodial staff supervisor are the other key participants. Both must support the program and must generate the same sense of commitment in their staff. A special O & M program will increase cleaning and maintenance work; staff dedication is necessary for an effective program.

Trained building inspectors also participate in all special O & M programs. These inspectors may be the ones who made the initial inspection for ACM. They may or may not be members of the in-house custodial or maintenance staff. In the O & M program, they will be inspecting the condition and other characteristics of the ACM.

Program Elements

Several aspects of a special O & M program are the same for all three types of ACM. For clarity and completeness, these steps are repeated in the description of each program.

SPECIAL PRACTICES FOR SPRAYED AND TROWELED-ON SURFACING MATERIALS

ACM that is sprayed or troweled on ceilings and walls is often the main source of airborne asbestos fibers in the building. Areas covered by ACM tend to be large. If the material is friable, fibers are slowly released as the material ages.

To reduce the level of released fibers and to guard against disturbing or damaging the ACM, the following measures should be taken.

Documentation, Education, and Training

The O & M program coordinator should:

- Record the exact location of ACM on building documents (plans, specifications, and drawings).
- Inform all building occupants and maintenance and custodial workers about the location of ACM and caution them against disturbing or damaging the ACM (e.g., by hanging plants or mobiles from the ceiling, or pushing furniture against walls). Be sure to give this information to new occupants and employees.
- Require all maintenance and custodial personnel to wear a half-face respirator with disposable cartridge filters or a more substantial respirator during the initial cleaning, and whenever they come in contact with ACM.
- Train custodial workers to clean properly and maintenance workers to handle ACM safely.

Initial Cleaning

Custodial staff should:

- Steam-clean all carpets throughout the building or vacuum them with a high-efficiency particulate air (HEPA)-filtered vacuum cleaner, but never with a

conventional vacuum cleaner. Spray vacuum cleaner bags with water before removal and discard in sealed plastic bags according to EPA regulations for removal and disposal of asbestos. Discard vacuum filters in a similar manner.

- All curtains and books should be HEPA-vacuumed. Discard vacuum bags and filters in sealed plastic bags according to EPA regulations for disposal of asbestos waste.
- Mop all noncarpeted floors with wet mops. Wipe all the shelves and other horizontal surfaces with damp cloths. Use a mist spray bottle to keep cloths damp. Discard cloths and mopheads in sealed plastic bags according to EPA regulations for disposal of asbestos waste.

Monthly Cleaning

Custodial staff should:

- Spray with water any debris found near surfacing ACM and place the debris in plastic bags using a dustpan. Rinse the pan with water in a utility sink. Report presence of debris immediately to the O & M program coordinator.
- Vacuum (HEPA) all carpets.
- Wet-mop all other floors and wipe all other horizontal surfaces with damp cloths.
- Dispose of all debris, filters, mopheads, and cloths in plastic bags according to EPA regulations for disposal of asbestos waste.

Building Maintenance

The special O & M program coordinator should ensure that recommended procedures and safety precautions will be followed before authorizing construction and maintenance work involving surfacing ACM. Specifically, containment barriers should be erected around the work area and workers should wear coveralls as well as respirators.

Maintenance staff should:

- Clear all construction, renovation, maintenance, or equipment repair work with the O & M program coordinator in advance.
- Avoid patching or repairing any damaged surfacing ACM until the ACM has been assessed by the asbestos program manager.
- Mist filters in a central air ventilation system with water from a spray bottle as the filters are removed. Place the filters in plastic bags and dispose of them according to EPA regulations.

Periodic Inspection

Building inspectors should:

- Inspect all ACM materials for damage or deterioration at least twice a year and report findings to the O & M program coordinator.
- Investigate the source of debris found by the custodial staff.

Custodial and maintenance staff should inform the O & M program coordinator when damage to ACM is observed or when debris is cleaned up.

An illustrated EPA pamphlet, "Asbestos in Buildings—Guidance for Service and Maintenance Personnel" (USEPA 1985a), may be especially useful in publicizing and initiating the special O & M program. Contact the Regional Asbestos Coodinator (RAC) or call the EPA toll-free line for copies of the pamphlet (see Appendix C for addresses).

The special O & M program should continue until all surfacing ACM is removed. Over time, the special O & M program may need to be altered if the ACM is enclosed or encapsulated.

SPECIAL PRACTICES FOR PIPE AND BOILER INSULATION

Asbestos-containing pipe and boiler insulation typically is a less significant source of airborne asbestos fibers than surfacing ACM. Unless damaged, protective jackets around such insulation prevent fiber release. Thus, the special O & M program for pipe and boiler insulation focuses on alerting workers to its location, inspecting the protective jacket (and pipe joints or elbows) for damage, and taking precaution prior to building construction activities. The program also includes repair and selected special cleaning practices.

Semiannual Cleaning

Custodial staff should:

- Spray with water any debris found near asbestos-containing insulation and place the debris in a plastic bag using a dustpan. Clean the pan with water in a utility sink. Report presence of debris immediately to the O & M program coordinator.
- HEPA-vacuum all carpets in rooms that have asbestos-containing insulation.
- Wet-mop all other floors and dust all other horizontal surfaces with damp cloths in rooms with asbestos-containing insulation.
- Seal all debris, vacuum bags, vacuum filters, cloths, and mopheads in plastic bags for disposal according to EPA regulations for asbestos waste.

Maintenance

The special O & M program coordinator should:

- Ensure that recommended procedures and safety precautions will be followed before authorizing construction and maintenance work involving pipe and boiler insulation. Specifically, containment barriers or bags should be positioned around the work area and workers should wear coveralls and respirators.

- Authorize repair of minor insulation damage with nonasbestos mastic, new protective jackets, and/or nonasbestos insulation following recommended repair techniques and precautions.
- Authorize large-scale abatement only after a complete assessment of the asbestos-containing insulation.

The maintenance staff should:

- Clear all construction, renovation, maintenance or equipment repair work with the O & M program coordinator in advance.
- Avoid patching and repair work on insulation until the ACM has been assessed by the asbestos program manager.

Periodic Inspection

Building inspectors should:

- Inspect all insulation for damage or deterioration at least twice a year and report findings to the O & M program coordinator.
- Investigate the source of debris found by the custodial staff.

Custodial and maintenance staff should inform the O & M program coordinator when damage to the insulation is observed or when debris is cleaned up.

The illustrated EPA pamphlet, "Asbestos in Buildings—Guidance for Service and Maintenance Personnel" (USEPA 1985a), may be useful for the special O & M program for pipe and boiler insulation. The O & M program should continue until all asbestos-containing insulation (including materials on pipe joints and elbows) is removed and replaced with another type of insulation.

SPECIAL PRACTICES FOR OTHER ACM

Most ACM that is neither surfacing material nor pipe and boiler insulation is hard and nonfriable. This type of ACM releases fibers only when manipulated (e.g., cut, drilled, sawed) or damaged. The special O & M program is designed to alert workers to the location of ACM and to avoid its disturbance or damage.

Documentation, Education, and Training

The O & M program coordinator should:

- Record the exact location of these types of ACM on building documents (plans, specifications, and drawings).
- Inform maintenance and custodial workers about the location of ACM and caution them about disturbance or damage.
- Train maintenance workers to handle ACM safely.[8]

4
The Building Survey

COLLECTING SAMPLES

Taking a sample of asbestos-containing materials (ACM) can damage the material and cause significant release of fibers. The following guidelines are designed to minimize both damage and fiber release.

- Wear at least a half-face respirator with disposable filters.
- Wet the surface of the material to be sampled with water from a spray bottle or place a plastic bag around the sampler.
- Sample with a reusable sampler such as a cork borer or a single-use sampler such as a glass vial.
- With a twisting motion, slowly push the sampler into the material. Be sure to penetrate any paint or protective coating and all the layers of the material.
- For reusable samplers, extract and eject the sample into a container. Wet-wipe the tube and plunger. For single-use samplers, extract, wet-wipe the exterior, and cap it.
- Label the container.

- Clean debris using wet towels and discard them in a plastic bag.
- For surfacing material, use latex paint or a sealant to cover the sample area. For pipe and boiler insulation, use a nonasbestos mastic.

SELECTING A QUALIFIED LABORATORY

The U.S. Environmental Protection Agency (EPA) runs a bulk asbestos sample quality assurance program. Updated lists of participating laboratories, their performance scores, and further information on the program are available from the EPA by calling the Toxic Assistance Office at (202)554-1404.

Checklist for Determining Contractor Qualifications

1. Contractors shall demonstrate reliability in performance of general contracting activities through the submission of a list of references of persons who can attest to the quality of work performed by the contractor.
2. Contractors must demonstrate ability to perform asbestos activities by submitting evidence of the successful completion of training courses covering asbestos abatement. At a minimum, the contractor shall furnish proof that employees have had instruction on the dangers of asbestos exposure, on respirator use, decontamination, and OSHA regulations.
3. Contractors must be able to demonstrate prior experience in performing previous abatement projects through the submission of a list of prior contracts, including: the names, addresses, and telephone numbers of building owners for whom the projects were performed. In rare circumstances inexperienced contractors may be qualified if they can demonstrate exceptional qualifications in the other contractor standards.

4. Additional evidence of the successful completion of prior abatement projects should be demonstrated by contractors through the submission of air-monitoring data, if any, taken during and after completion of previous projects in accordance with 29 CFR 1910.1001 (e).

5. Contractors must possess written standard operating procedures and employee protection plans which include specific reference to OSHA medical monitoring and respirator training programs. In addition, the contractor must be prepared to make available for viewing at the job site a copy of OSHA regulations 29 CFR 1910.1001 governing asbestos controls, and Environmental Protection Agency regulations 40 CFR Part 61, Subpart M (NESHAPS), governing asbestos stripping work practices, and disposal of asbestos waste.

6. In those states which have contractor certification programs, contractors must possess state certification for the performance of asbestos abatement projects.

7. Contractors must be able to provide a description of any asbestos abatement projects which have been prematurely terminated, including the circumstances surrounding the termination.

8. Contractors must provide a list of any contractual penalties which the contractor has paid for breach of or noncompliance with contract specifications, such as overruns of completion time or liquidated damages.

9. Any citations levied against the contractor by any federal, state, or local government agencies for violations related to asbestos abatement shall be identified by contractors, including the name or location of the project, the date(s), and how the allegations were resolved.

10. Contractors must submit a description detailing all legal proceedings, lawsuits, or claims which have been filed or levied against the contractor or any of his past or present employees for asbestos-related activities.

11. Contractors must supply a list of equipment that they have available for asbestos work. This should include negative air machines, Type C supplied air systems, scaffolding, decontamination facilities, disposable clothing, etc.[9]

Definition and Description of Factors for Assessing the Need for Corrective Action

Condition of the ACM (What the Contractor Should Look For)

Factor: Deterioration or Delamination and Physical Damage. An assessment of the condition should evaluate: (1) the quality of the installation, (2) the adhesion of the friable material to the underlying substrate, (3) deterioration, and (4) damage from vandalism or any other cause. Evidence of debris on horizontal surfaces, hanging material, dislodged chunks, scrapings, indentations, or cracking are indicators of poor material condition.

Accidental or deliberate physical contact with the friable material can result in damage. Inspectors should look for any evidence that the ACM has been disturbed: finger marks in the material, graffiti, pieces dislodged or missing, scrape marks from movable equipment or furniture, or accumulation of the friable material on floors, shelves, or other horizontal surfaces.

Asbestos-containing material may deteriorate as a result of either the quality of the installation or environmental factors which affect the cohesive strength of the ACM or the strength of the adhesion to the substrate. Deterioration can result in the accumulation of dust on the surface of the ACM, delamination of the material (i.e., separating into layers), or an adhesive failure of the material where it pulls away from the substrate and either hangs loosely or falls to the floor and exposes the substrate. Inspectors should touch the ACM and determine if dust is released when the material is lightly brushed or rubbed.

If the coated surface "gives" when slight hand pressure is applied, or the material moves up and down with light pushing, the ACM is no longer tightly bonded to its substrate.

Factor: Water Damage. Water damage is usually caused by roof leaks, particularly in buildings with flat roofs or a concrete slab and steel beam construction. Skylights can also be significant sources of leaks. Water damage can also result from plumbing leaks and water or high humidity in the vicinity of pools, locker rooms, and lavatories.

Water can dislodge, delaminate, or disturb friable ACM that is otherwise in good condition and can increase the potential for fiber release by dissolving and washing out the binders in the material. Materials which were not considered friable may become friable after water has dissolved and leached out the binders. Water can also act as a slurry to carry fibers to other areas where evaporation will leave a collection of fibers that can become suspended in the air.

Inspect the area for visible signs of water damage, such as discoloration of, or stains on, the ACM; stains on adjacent wall or floors; buckling of the walls or floors; or areas where pieces of the ACM have separated into layers or fallen down, thereby exposing the substrate.

Close inspection is required. In many areas, staining may occur only in a limited area while water damage causing delamination may have occurred in a much larger area. In addition, the water damage may have occurred since the original inspection for friable material, causing new areas to become friable and require a reinspection.

Delamination is particularly a problem in areas where the substrate is a very smooth concrete slab. Check to see if the material "gives" when pressure is applied from underneath.

Potential for Disturbance or Erosion

Factor: Air Plenum or Direct Airstream. An air plenum exists when the return (or, in rare cases, conditioned) air leaves a room or hall through vents in a suspended ceiling, and travels at low speed and pressure through the space between the actual ceiling and the suspended ceiling or ducts. The moving air may erode any ACM in the plenum. In evaluating whether an air plenum or direct airstream is present, the inspector must look for evidence of ducts or cavities used to convey air to and from heating or cooling equipment, or the presence of air vents or outlets which blow air directly onto friable material.

A typical construction technique is to use the space between a suspended ceiling and the actual ceiling as a return air plenum. In many cases, the tiles in the suspended ceiling must be lifted to check whether this is the case. Inspection of the air-handling or HVAC equipment rooms may also provide evidence (such as accumulated fibers) of the presence of this material in the plenums.

Special attention should be paid to whether frequent activities (such as maintenance) disturb the material in the plenum. It is also important to check for evidence that the material is being released or eroded (i.e., whether it has deteriorated or been damaged so that the material is free to circulate in the airstream).

Factor: Activity. In general, for a site to show a high potential for disturbance, it must be exposed (visible) and accessible, and be located near movement corridors or subject to vibration.

The amount of ACM exposed to the area occupied by people will determine the likelihood that the material may be disturbed and thereby cause the fibers to freely move through the area. An ACM is considered exposed if it is not well contained. For a material not to be exposed, a physical barrier must be complete, undamaged, and unlikely to be removed or dislodged. An ACM should be considered exposed if it is visible.

Height above the floor is one measure of accessibility. However, objects have been observed embedded in ceilings 25 ft or more in height. Nearness of the friable ACM to heating, ventilation, lighting, and plumbing systems requiring maintenance or repair may increase the material's accessibility.

If the ACM is located behind a suspended ceiling with movable tiles, a close inspection must be made to determine the condition of the suspended ceiling, the likelihood and frequency of access into the suspended ceiling, and whether the suspended ceiling forms a complete barrier or is only partially concealing the material. Asbestos-containing material above a suspended ceiling is considered exposed if the space above the suspended ceiling is an air plenum. Suspended ceilings with numerous louvers, grids, or other open spaces should be considered exposed. There are numerous buildings in which areas above ceilings serve as the return air duct for a heating system.

If friable ACM can be reached by building users or maintenance people, either directly or by impact from objects used in the area, it is accessible and subject to accidental or intentional contact and damage.

In addition, the activities and behavior of persons using the building should be included in the assessment of whether the material is accessible. For example, persons involved in athletic activities may accidentally damage the material on the walls and ceiling of gymnasiums with balls or athletic equipment. To become fully aware of occupants' use of the building, the inspector should consult with building staff or personnel.

When assessing activity levels, consider not only the movement caused by the activities of people but also movement from other sources such as high vibration from mechanical equipment, highways, and airplanes. Another source of vibration is sound, such as music and noise, which sets airwaves in motion at certain frequencies. As these sound waves impact on ACM, they may vibrate the material and contribute to fiber release; therefore, more fibers may be released in a music practice room or auditorium than in the rest of the building.

Factor: Change in Building Use. A planned change in the use of the building from, for example, a primary to a senior high school may imply significant changes in the potential for erosion or disturbance. Of particular note is the increased potential for damage from balls to previously inaccessible ceilings in gymnasiums. The addition of machinery (such as dust collectors in wood or metal shops) to a school or office building may introduce vibrations which, again, may be a future cause of concern. The inspector should exercise judgment and draw on experience in evaluating the likely effect of such changes.

BASIC CONSTRUCTION OF SCHOOL AND COMMERCIAL BUILDINGS [10]

To understand the layout of a building an understanding of the components and systems is necessary.

In multistory buildings, generally a utility core runs vertically through the building. From this core, service runs branch to individual floors. Elevators are generally bundled, and stair towers run vertically through the structure.

Mechanical Systems

Mechanical systems are those systems designed by the mechanical engineer. They include the HVAC system, the plumbing system, and in the eastern regions of the U.S., elevators.

Heating, Ventilating, and Air Conditioning (HVAC) Systems

Individual spaces or zones in a building are served by supply and return air and a thermostat to activate the HVAC system. The supply and return may be in ductwork or in a plenum. Plenums are spaces; for example, the space above a dropped acoustical ceiling and below the roof or floor above. In most cases, the plenum is used for return air, that is, the air leaves the room, enters the plenum, and is drawn into the mechanical room from the plenum.

All HVAC systems have a method of heat transfer. The heat transfer may occur in the central mechanical space or "plant" but in large buildings or complexes it will occur in individual mechanical rooms. From the mechanical room the conditioned air is sent to individual rooms within the building, and the return air is carried back to the mechanical room where it is filtered and reconditioned. In addition, some make-up air is added to provide a source of fresh, outside air to the building.

Air Systems

There are two types of air HVAC systems—single duct and double duct. (This refers to supply only; return is accomplished in yet another duct.) A single duct system delivers either heated or cooled air at a constant temperature from the air conditioning equipment through ductwork.

When a double duct system is used, one duct carries cooled air while the other carries heated air. The two ducts meet at a mixing box, where the amount of heated or cooled air is regulated.

Water Systems for HVAC

Heated and/or cooled water is delivered to a fan coil unit, where the air is introduced. Air is blown across coils as regulated by dampers, again activated by the requirements of the individual space. This air is introduced through a separate duct system from the mechanical or fan room, or from direct connection to the outside.

Refrigerant Systems

These are packaged units that supply heating or cooling directly to a space through a wall or roof. In general, these are used only in specialized installations in commercial buildings.

Radiant Systems

Radiant systems include any number of devices which are either embedded in the wall or floor, or are set as radiators, usually along an exterior wall. They are usually used for heating and function by radiating heat directly into a space.

Since the primary function of the HVAC system is to heat and cool building spaces, insulation is used to inhibit unwanted heat transfer. Insulation is typically found on the outside of boilers and on the breeching or flue which conveys waste gases from the combustion process. Blanket or batt insulation is sometimes found inside ducts, and insulation is sometimes sprayed on the outside of ducts. Each of these types of insulation should be considered suspect ACM. In addition, gasket material on boiler doors, rope used as filler in openings, and vibration-dampening cloth connecting sections of ductwork may contain asbestos.

HVAC systems which use chilled water will often include a cooling tower where excess heat is rejected to outdoor air. Cooling tower baffles and sometimes filter media (fill) are constructed with ACM; the slats are frequently Transite, which is a type of ACM.

Plumbing Systems

Plumbing systems include any water, gas, or other fluid which is piped through a building, and in some cases disposed of as waste. Also considered part of the plumbing system is air when used in a non-HVAC manner, such as compressed air in factories. Plumbing systems consist of piping (horizontal pipes are called runs, vertical pipes are called risers).

The water systems in a building are of four types: consumed, circulated, static, and controlled. The consumed system is potable water for use and consumption by building occupants. Static water is water used for the fire protection, and controlled water is water used to maintain relative humidity within the building. The drawings or plans may use notations such as CW, PW, or others, for chilled water and potable water.

The use of asbestos in plumbing systems is usually for the purpose of temperature control. Generally it can be found on the piping and equipment which heat water and/or maintain water at a stable temperature. Insulating materials prevent heat loss from hot pipes and equipment and water condensation on the outside of cold pipes and equipment. ACM is typically limited to consumed and circulated water systems where some temperature control is needed.

In addition to plumbing insulation, asbestos cement pipe may have been used in the plumbing system. The pipe is concrete-like in appearance and is known by the tradename "Transite."

ELECTRICAL SYSTEMS

Electrical systems within a building may appear very complex, but are simple in their basic design. Each building includes an electric service entrance; the point where the energy enters the building.

Asbestos use in electrical systems has included:

- Transite ducts for electrical cable runs
- partitions in electrical panels
- asbestos cloth to bind bare cables
- insulation on stage lighting and on the wires to those lights
- insulation behind bulbs

Of great concern in inspecting electrical systems is the potential hazard to the inspector from unsafe inspection procedures.

- Whenever possible, conduct the inspection accompanied by a building operator who is familiar with the electrical equipment, its operation and location.
- Beware of exposed electrical wires and components.
- Do not use a wetting solution near an electrical system.
- When taking samples of surfacing or other suspect materials be careful not to penetrate to electrical components that may be located underneath or behind.

CONTRACT DOCUMENTS

Contract documents or construction documents are the legally binding drawings and specifications which are used to construct the building. They consist of:

- working drawings, called plans
- specifications, called specs

- other documents, including addenda, change orders, shop drawings, submittals, and as-builts

These documents are a rich source of information for building inspectors. Building owners should be able to provide copies. If they are not available, check with the local building permit agency. If the jurisdiction has a plan check and building permit procedure, the plan check agency likely retained a set of the drawings.

The plans are a set of drawings which indicate the finished appearance and construction of the building. They are not a set of exact instructions for the contractor.

A title block will appear along the right side or in the lower right-hand corner of each sheet of the set of drawings. When beginning your review of the drawings, carefully examine the title block for the following information:

- the name of the architecture or engineering firm
- the date of the drawings
- the sheet numbers
- the project number

The sheet numbering system for the entire set of drawings reflects the manner in which the drawings were prepared. Just as the design of the building is a collaborative effort of an architect and engineers, the drawings and specifications are prepared by each of these professionals. Altogether, a complete set of drawings will likely include:

- mechanical (HVAC)
- structural
- plumbing
- architectural
- electrical

When examining the numbering of the drawings, one will find that the drawings are divided by discipline. That is, the architectural drawings are together, the structural drawings are together, and so on. The numbering is then dependent on the discipline. Architectural drawings will be identified with an A, the mechanical (HVAC only) with an M, the plumbing with a P, and the electrical with an E.

There is little standardization in the production of drawings, and thus no set of rules can be given for the way each architect or engineer prepares not only a set of drawings, but also individual items within that set.

Drawings can be divided into several generic types:

- Plans—drawings of the building as viewed from above, these include floor plans, foundation plans, framing plans, roof plans, and electrical plans, and should not be confused with the entire set of drawings which is also referred to as the set of plans
- Elevations—drawings of the building as viewed horizontally from outside
- Sections—drawings cut (vertically) through the building or building parts
- Details—expanded views of small areas that can be drawn in plan, elevation, or section
- Notes—comments and explanations
- Schedules—a tabular display of information

In reviewing the drawings, be sure to check for a list of symbols. Each building material in a set of drawings is depicted, when cut in section, by a material indication. If a legend appears in the set, use it as a guide.

A drawing reference that may be encountered on a set of building plans is a "revision." It is depicted by a triangle around a number; a portion of the drawing itself may also be clouded to further indicate where the revision applies. The number identifies the revision. The key to the numbering is found in the area adjacent to the title block.

Mechanical Drawings

The mechanical engineer prepares drawings for both the HVAC system and the plumbing system. Mechanical drawings consist of mechanical plans, which are based on the building's floor plans. They indicate the routing of ductwork and piping systems necessary for HVAC, as well as details, notes, schedules, sections, and elevations. Mechanical plans may include a system schematic, or flow diagram, to indicate how the HVAC system operates.

When reviewing the mechanical drawings, the inspector needs to become familiar with the kind of HVAC system used, and the location of the various parts of the system. It is necessary to verify information obtained from these drawings by field inspection.

Structural Drawings

Structural drawings will consist of foundation plans, floor framing plans, roof framing plans, structural elevations, details, notes, and schedules.

Many buildings use a structural grid based on column location referenced by numbers or letters. The grid provides a way to organize the building and to communicate about specific areas.

If the building has fireproofing, it may not be indicated on the structural drawings as it is a finish or surfacing applied to the skeleton, not part of the skeleton. Thus it is the architect's responsibility, not the structural engineer's.

However, to understand where the fireproofing has been applied, where the beams are located that it is applied to, and the amount of area covered, the inspector will need to examine the structural drawings.

Structural notes will often include a building code reference. These codes identify the name and the year of the official building code which governed the design of the structural elements. This reference can be an invaluable tool. Building codes in effect when the building was erected may have specified fireproofing and other materials which are likely to contain asbestos.

Plumbing Drawings

Plumbing drawings include plumbing plans, which are based on floor plans, notes, schedules, riser diagrams, and other required supporting drawings. These are often the most important drawings for an ACM inspection.

Architectural Drawings

Architectural drawings show finished surfaces and materials. Of note is the floor plan, which is cut through the building at about four feet above the floor. The floor plan is the basis for the mechanical, plumbing, and electrical drawings.

Another important drawing is the demolition plan, which represents those portions of the building which will be demolished as a part of a renovation project.

Commercial buildings often have repetitive units, e.g., rooms, doors, and windows. To organize these spaces, schedules are developed to identify and describe specific rooms, doors, and windows. The specific item is referenced to the schedule with a symbol as shown on the legend in the set of drawings. If a legend does not accompany the drawings, use extreme care when working with room, door, and window designations as the same number may be used repeatedly, and the only difference will be in the symbol in which the number is lettered.

Be aware of differences between the room numbering scheme in the plans and the current numbering of the rooms in the building. It may be necessary to cross-list the numbers to equate the design information with the information determined from on-site investigation.

When reviewing drawings, the intention should be to familiarize one's self with the layout of the building, and then examine in detail the finishes or details at the exterior wall and other areas where one suspects ACM may be found.

Often, when referenced on the drawings, a material will be listed with the notation "OR EQUAL." This notation allows for the contractor to make a substitution of another equivalent material.

Electrical Drawings

Electrical drawings consist of the floor plan-based power and lighting plans, notes, schedules, details (if required), and calculations to support the load requirements. A cursory review of the electrical drawings is normally all that is required to familiarize one's self with the location of equipment and equipment rooms. Electrical drawings are largely schematic. The exact location of all items, excepting panels, lighting, switches, and receptacles, is determined in the field, and as such, needs to be verified.

Specifications

The specifications (specs) for a project are a written set of standards and procedures which inform the contractor of what materials and standards are necessary for the successful completion of the facility. The specs are generally in book form and accompany the drawings as a portion of the contract documents. It is not unusual for buildings to have several items included which are not noted in the specs. Changes during construction, maintenance activities, renovations, or other occurrences may often not be shown on plans or noted in the specs.

5
The Contractor's View of a Project

ASSESSING THE WORK AREA

An important rule of thumb for any asbestos abatement contractor is never to accept, or bid, a project, without first viewing and assessing the site. There is much valuable information to be gained during one of these assessments, such as determining the size of the job (number of square feet of asbestos-containing material [ACM]), or examining the configuration of the ceiling surface (irregular ceiling shape can increase the amount of ACM originally believed to be present). A survey such as this also provides a basis upon which the contractor can formulate an effective strategy for asbestos removal and/or control.

Check Analytical Results of Bulk Samples

Probably the first questions that a contractor should ask during the pre-bid walk-through survey are: Who did the initial survey to identify the asbestos, what type of sampling was conducted, and what forms of analysis were used. The contractor should ensure that appropriate bulk sampling was performed by qualified individuals using proper analytical methods. A laboratory that participates in the EPA bulk asbestos identification quality assurance program is a minimal requirement (accreditation by the American Industrial Hygiene Association is also preferred). The contractor should then review the analytical results of the bulk samples to determine the types and percentages of asbestos present.

There are several reasons why this type of information will be of benefit to the contractor. First, the analytical reports provide excellent documentation that can be used in establishing a project file. This file can then be used as a good source of reference, should any questions arise concerning the ACM in the building. Information contained in the analytical reports is also important because different types of asbestos will require various handling techniques. For instance, amosite is considered by some scientists to be more hazardous than chrysotile, in addition to not accepting wetting agents as well, and will require different handling procedures. Fiber counts will usually be much higher when handling amosite as opposed to chrysotile. If analytical reports are not available prior to, or during, the survey, the contractor should obtain his/her own by including it as part of the assessment. It is important that the information from these reports be used as the main criteria on which to base decisions, rather than word-of-mouth from a resident maintenance worker or other building occupant which could lead to confused facts or other misinformation.

Inspect the Nature of the ACM

The contractor should determine the hardness and texture of the ACM to be removed by touching it. He/she should also note whether or not it has been painted over. (Note: A high efficiency cartridge-type respirator should be worn when conducting these tests.) The contractor may also wish to test a sample area of ACM to determine its ability to absorb amended water. This can be done by using a plant sprayer. If the material cannot absorb wetting agents, other appropriate strategies will need to be developed which may increase the cost and project time.

Check Accessibility of Material

Note the accessibility of all materials for removal; that is, whether or not the ACM is accessible enough to remove. If not, an alternative means of control might have to be used, such as encapsulation or enclosure. Several factors that may enter into this determination are ceiling height, false ceilings, pipes, sprinklers, ducts, sloping floors, fixed barriers, etc.

Check for Difficulty of Isolating the Work Area

Another important concern is isolating the area in which removal will take place. Is it possible to enclose the area completely by using 6 mil polyethylene? Or will other measures have to be implemented in certain areas to adequately isolate the removal site? In cases such as school buildings, it may be easiest simply to line the walls and floors with two layers of 6 mil polyethylene, since the contractor will usually remove all desks and chairs from the work area. However, in cases such as a church or computer room, plywood and plastic enclosures may have to be constructed so that the materials left in the room will not be contaminated by the asbestos removal activities, or damaged by water. Other sections of this book, "Preparing the Work Area," and "Establishing a Decontamination Unit," Chapter 7, further discuss these practices.

Determine If Areas Adjacent to Abatement Activity Will Be Occupied

If areas adjacent to the abatement activity will remain occupied, several important practices should be observed. Most importantly, the HVAC system will need to be altered, or the opening of the duct into the work area should be completely sealed off. This sealing of the HVAC helps ensure that airborne fibers will not be drawn into the air return system and dispersed throughout adjacent areas, or that the supply system will not place the work area under positive pressure and cause airborne fibers to escape. To provide documentation that contamination of adjacent areas has not occurred, a qualified person should take background air samples in each of the areas before abatement work begins. These results are then compared to the results of samples taken in these areas during and after the work is completed. By doing this sampling, it can be demonstrated that other areas were not contaminated as a result of the asbestos abatement work.

Determine Room Volume and Natural Air Movement in the Work Area

During this walk-through survey, consideration should be given to the number and placement of negative air units. An estimate of the air volume in the work area is necessary for determining the number of units needed to achieve the desired number of air changes per hour. Also, the way in which air will move through the work area is a consideration in placement of the negative air units.

Check Items Requiring Special Protection

During the pre-bid walk-through, items requiring special protection should be noted. These items might include walnut paneling, trophy cabinets, glass piping, carpets, lab equipment, dangerous chemicals, computers, and elevators. In the case of walnut paneling, common sense should be used when hanging polyethylene to enclose the work area. Care must be used when tacking up the plastic so that the paneling will not become damaged. The nails should be placed between the panel strips in the natural gaps as near the ceiling as possible to prevent any small holes from being visible.

For trophy cabinets that are stationary and must remain in the work area while removal is taking place, proper measures must be taken to ensure that the cabinet is adequately enclosed with 6 mil polyethylene. During this initial survey, the contractor should note the condition of any of these cabinets and the exact contents of each to prevent any future conflicts that could result if someone were to claim that something was damaged or missing.

Glass piping is another item that the contractor should note during the pre-bid walk-through, since special procedures must be followed to ensure that it does not become damaged. These glass/ceramic pipes will often contain hazardous materials (acids, hazardous waste, etc.); therefore, the pipes should be tagged and/or labeled as containing hazardous materials, and workers should avoid contacting them if possible.

These glass pipes are often found in the vicinity of other pipes which have asbestos-containing lagging on them; therefore, contingency procedures must be established to prevent and handle hazards which could develop from working around these pipes.

Determine If Existing Carpet Is to Be Removed

Special note should also be made of where carpeting is located in the facility. In most cases, the carpeting should be removed completely from the area in which the asbestos removal will be taking place. When fibers settle on a carpeted surface, they often penetrate through to the floor and become trapped underneath. Once this occurs, repeated traffic over the area will cause the fibers to be redispersed throughout the surrounding air. If the carpet is specified for removal, assess the difficulty of removing it (e.g., the carpet may be glued in place). Also, consideration must be given to disposal requirements and procedures.

Note Any Materials or Equipment Which Will Require Special Handling

Lab equipment and/or dangerous chemicals should be examined closely by the contractor during the pre-bid walk-through survey. It may be necessary to remove much of the equipment and/or chemicals from the work area before abatement activities take place. If the contractor's employees will be moving expensive lab equipment or chemicals, the contractor should ensure that all items are appropriately handled through training and/or direct supervision. This may be a tedious process requiring extra time to complete. In some cases, the building owner may have his/her own maintenance personnel perform these functions before the contractor comes in to begin work.

Note Stationary Objects That Require Special Attention

As previously mentioned, if the abatement work area will be in a room that contains computers which cannot be moved, other strategies must be developed such as building an elevated platform (plywood and plastic) over the terminals.

Elevators can also be a major problem on an asbestos abatement job. The elevator, or the shaft, can become contaminated with ACM, or their movement can cause air displacement in contaminated areas. The contractor will need to take special precautions to properly seal off the door with 6 mil polyethylene (even plywood in some cases), and to key the elevators not to stop at the floor(s) on which the work area is located.

Other Considerations

To prevent any misunderstandings or conflicts, it is imperative that the job specifications spell out exactly who is to pay for the utilities used during the project. Usually, the building owner will pay these expenses, but if not, this should be clearly understood by both sides before work begins. Likewise, the waste water filtration and disposal method should be agreed upon and be specified.

The contractor should also document all preexisting damages in the areas in which his/her employees will be working. Photographs, videotapes, diagrams, lists, and tape recordings may be used for these purposes. This documentation should include all surface damages (walls, tables, desks, etc.), vandalism, roof leaks, or other water damage. This consideration is important because often after a project has been completed, the building owner, or another facility operator, will claim that some damages occurred as a result of the contractor's work. By utilizing the list that was developed at the beginning of the project, the contractor can verify whether the damages were preexisting, and not a result of the contractor's work.

Other important aspects that should be considered by a contractor when conducting a pre-bid walk-through survey include an estimate of the temperature when the project is scheduled to begin. It may be that the bid is at the end of the summer, and the project is scheduled to begin in the winter, or vice versa. In these cases, appropriate climate control strategies will need to be implemented. Also, at this time it should be decided who will provide security at night or during off-hours to assure that no unauthorized entries into the contaminated work area will occur. Additional safety hazards that need to be considered include all electrical circuits and/or receptacles, equipment, etc. Since the work area in an asbestos abatement job will commonly contain large amounts of water, the potential for electrical hazards will be greatly increased. During the pre-bid walk-through, the contractor should make note of all these potential hazards. Once the building owner is made aware of these situations, an appropriate plan of action can be implemented. It may be possible and appropriate to shut down all power to the work area while the project is going on. If not, other precautions will need to be taken.

Consideration must also be given as to where the contractor will be able to park vehicles or trailers. Are there adequate facilities presently available, or will other arrangements have to be made? Along with this, consideration must be given to where the contractor's equipment and supplies will be stored. If there is not adequate space available on the job site, it may be necessary to rent additional space at some nearby location. Care must be used so that the rented space will not become contaminated. (Note: Lining the space with two layers of 6 mil polyethylene is recommended.)

Possibly the most important aspect to consider during the pre-bid survey is whether full or partial removal will take place. If partial removal will be performed, the airborne fiber clearance levels in the contract specifications should be examined closely to determine if that level is achievable.

Another area of concern during the walk-through should be the configuration of the walls and surfaces for attaching tape. This is important in determining how the polyethylene sheeting will have to be hung to adequately enclose the work area. Care must be used when hanging polyethylene so the walls will not be damaged, but the plastic will remain in place until intentionally moved. This is an important consideration at this time, since the contractor will have to estimate how much material will be needed to enclose all work areas. Often the building owner may want the project to be inconspicuous to the general public.

For this reason, opaque polyethylene may have to be used to construct tunnels from the work areas outside to the waste disposal trucks. Additionally, depending on the nature of the work area, special tools, equipment, and manlifts or scissorlifts may have to be utilized during prepping of the work area.

The location and type of decontamination units should also be a major consideration before submitting a bid. Will it be possible to have one central decontamination unit, or will it be necessary to establish multiple stations? Some contractors may have their own units (i.e., trailers), but many choose to build them on-site. Many buildings in which asbestos removal takes place already contain shower facilities (i.e., school buildings, gymnasiums, etc.). Under no circumstances should the contractor ever permit his/her employees to use these as part of the decontamination sequence. Separate facilities should be constructed utilizing appropriate waste water filtration equipment. An advantage of building temporary site units is that the chance of residual contamination is reduced, since they will be demolished at the end of the project and disposed of.

Also, a major area of concern when assessing a facility prior to beginning work is identification of any hot surfaces (pipes) that could present a hazard to abatement workers. First, it should be noted whether the pipes will be active or inactive. If they are active, appropriate measures will have to be taken to ensure that workers will not contact these surfaces. If the lines are inactive, work may be carried out as it would on any other surface of normal temperature. The contractor should investigate the types of reinsulation that will be required on surfaces and pipes after the ACM has been removed. The original material was there for some specific purpose; thus, replacement material with similar properties will probably be necessary.

If Type C air-supplied respirators will be used, the contractor must determine whether or not the hoses will reach the work area from the air-generating source. Low pressure air-supply lines cannot exceed 300 ft, according to OSHA regulations.

Another important aspect that must be considered by a contractor before bidding a project is who will pay for the air monitoring, and whether or not the person conducting the monitoring is qualified. This should be established in the specifications. The building owner should always be responsible for the daily air sampling, but the contractor is often responsible for (and required by OSHA) conducting personal air sampling on the asbestos abatement employees.

The contractor should ensure that the job specifications allow adequate time for their company to complete the job with a high degree of quality. If specifications call for a hurry-up job, the contractor should inform the building owner or architect if they do not feel that adequate time is available to complete the project. Attempting to perform the job hastily may only result in sloppy work and may needlessly endanger the health and safety of employees or other building occupants.

Finally, since there are an increasing number of asbestos abatement projects being undertaken these days, often the people directly involved with attempting to coordinate an asbestos abatement program for a facility may not be adequately educated in what needs to be included in the job specifications. Inevitably, there will be cases in which specifications from other projects are photocopied and sent out for bids. These are often not at all applicable to the particular facility of concern. The contractor should ensure that the specifications he/she is bidding on are designed for the work and work area of that facility. Though other specifications can usually serve as a guideline for developing a new set of specifications, they should never be used verbatim from one project to another. No matter how similar projects may seem, each one is different in some way.

These are not all of the special considerations that need to be examined when conducting a pre-bid walk-through survey of an asbestos-containing facility; rather, they are some common concerns that typically should be investigated before beginning any asbestos abatement project. These aspects are important because they could cost the contractor's company a substantial amount of time and money, in addition to possibly endangering the lives of employees or other building occupants. It is imperative that the contractor and the building owner have a firm understanding as to exactly how each step of the project will be carried out.[11]

DESIGN AND USE OF A PROJECT LOGBOOK

Prior to the start of any asbestos abatement project, a logbook should be established. The logbook serves as a vehicle for maintaining all the records associated with a project. At a minimum, included in the book should be copies of the employees' medical reports, copies of any accident or injury reports, air sampling results, notes concerning any deviation from standard work procedures, sign-in sheets, and all other pertinent documents, permits, correspondence, photographs, or records. Many of these records will be duplicated elsewhere, such as medical records in the employees' personal files, etc.

The logbook serves many important functions. It provides a ready reference for each project that can be presented at any time during the project, or long after its completion. It may be produced by the contractor to demonstrate to future clients the procedures followed during a project. The logbook can be an important tool for planning future jobs and estimating costs. When planning a project similar in nature, it can aid in estimating how long the project will take to complete, how many people will be necessary, and how to approach specific problems.

The following examples may help to illustrate this point. Example: Prior to the start of one job, the contractor probed the depth of the fireproofing and found it to be approximately 3 in. deep. During the removal, they discovered that approximately 50% of the fireproofing was over 6 in. deep, and in some areas, 9 in. deep. A few extra minutes of probing the depth of the fireproofing would have saved much time and money during this project. Unfortunately, this was not the first time that this had happened to this company. Had it been recorded in a logbook the first time it occurred, and changes in the standard procedures for estimating the amount of material made, this problem would probably have been avoided.

Example: A removal project of 24,000 ft^2 was two days ahead of schedule, with only the sprayback of treated cellulose remaining to be completed. Three days after this sprayback material was applied, it began to fall from the ceiling. It took the contractor an extra week of work to remove this material and replace it with a different substance. The problem appears to have resulted from the inability of the material to adhere to the substrate, since temperatures during application exceeded 95°F in some cases. Notes of this problem were maintained in the project logbook and corresponding instructions added to the standard operating procedures to prevent this from occurring again.

A project logbook may help in protecting a contractor from future liability concerning a specific project. A logbook indicates that the contractor performing the work actually attempts to do the best job possible using state-of-the-art techniques. The sign-in sheets maintain a record of all people entering and exiting the work area; for what purpose; for how long; and what personal protective equipment they need. This information, coupled with the air-sampling data, can quickly be used to estimate how much asbestos the person was exposed to, and for how long. Copies of daily inspection reports will also reveal if employees were wearing the appropriate protective equipment, and whether or not it was adequate in protecting them from the airborne fiber levels documented by the air-sampling results. This information would be very valuable if needed for litigation in the future. It is important to note that all records must be kept, not just a portion of them. Example: The following is a hypothetical example. The year is 1996; a woman dies of lung cancer. Her husband recalls that when she worked in a building 20 years before, the owner had stripped asbestos-containing fireproofing from the boiler room. A suit is filed against the building owners and the contractor who performed the removal work. Although the contractor performed air sampling throughout the projects, no records were kept regarding work practices; other people in the area; whether the air-handling system was on or off; or where the waste was disposed of. Since this is a hypothetical case, speculation on the outcome would not be appropriate. However, the contractor would have a better defense if proper records were maintained.

The logbook should be well organized, but in a style decided by the contractor. There are two common methods of organization. First, there is the day-by-day method, such as a ship captain's log. If this method is chosen, a looseleaf or bound notebook with dividers labeled with each day should be maintained for each job. Be sure to make entries on days during which no work is done, including how the integrity of the jobsite was maintained.

Another, more common, method of organizing a logbook is by activity. Using this method, a looseleaf notebook is divided into each activity and all documentation, notes, and receipts concerning that activity are maintained in the appropriate section. The following outline is one which could be used in organizing a logbook. It should be noted that this is just one outline; depending on the requirements of each project, some sections may not apply, while additional ones may be necessary.

SECTION	CONTENTS
Prework	EPA or state notification forms, any necessary state papers, licenses, county or city permits (contractor license, disposal permits, etc.). Records regarding the bonding company, size of bond, insurance coverage, etc.
Contract	Contract specifications, including all drawings/ diagrams. Specifications would be in this section.
Personnel	Personnel records, including employment applications, W-4 withholding forms, medical records, and any other records pertaining to each employee. Some firms also have their employees sign certificates stating that they have read and understand the OSHA asbestos standard CFR 1910.1001, have been trained in asbestos removal techniques, trained, and fit-tested for respirators, etc.

Sign-In	A separate section containing the daily sign-in sheets indicating when each employee went in and out of the work area, their affiliation, and their purpose for entering the work area. In this section would be a list of all personnel authorized to enter the contaminated area. Also in this section is a record of each employee's work hours for payroll purposes.
Subcontractors	This section contains a record of all subcontractors' activities, including copies of the contract, names, dates, etc.
Air Monitoring	All air sampling for the project should be included in this section. Area air-sampling and personal-sampling results should be presented. Also presented in this section should be a copy of the sampling and analytical method used, along with information concerning who performed the work. Waste Disposal Records of waste disposal activities, including trip tickets, should be kept in this section.
Daily	Copies of daily inspection reports should be maintained. It is important to include comments on unusual aspects of the project, to address any problems that arose, and to indicate how they were handled.
Other Sections	Other sections may be added as necessary (possibly injury/illness reports, receipts for rental equipment, lodging, outside inspections, newspaper clippings, etc.).

The responsibility for maintaining the logbook should be assigned to responsible personnel who can handle a court appearance, should this ever become necessary. Normally, this function is performed by the jobsite supervisor, site engineer, or industrial hygienist. Upon conclusion of the job, this person may write a one-page summary of the project. This summary can then be compiled with others and produced as evidence of previous jobs performed by the contractor, for prospective clients.[12]

6
Sample Specifications

This chapter contains sample specifications used by agencies experienced in asbestos projects. A careful review of these samples will provide an excellent basis for the preparation of new contract documents for most removal projects.

SAMPLE ASBESTOS ABATEMENT SPECIFICATIONS

DEPARTMENT OF THE NAVY NAVAL FACILITIES ENGINEERING COMMAND GUIDE SPECIFICATION[13]

NFGS-02075N (January 1987) Use in lieu of NFGS-02075 (February 1982), Amendment -1 (April 1984) and Amendment ND-1 (February, 1985)

Removal and Disposal of Asbestos Materials

Table of Contents

1.1.1 Military Standard (Mil. Std.)
1.1.2 Federal Standard (Fed. Std.)
1.1.3 Code of Federal Regulations (CFR) Publications
1.1.4 American National Standard Institute (ANSI) Publications

1.2 REMOVAL AND DISPOSAL

1.2.1 Description of Work
1.2.2 Definitions
1.2.3 Title to Materials
1.2.4 Protection of Existing Work to Remain
1.2.5 Medical Requirements
1.2.6 Training
1.2.7 Permits and Notifications
1.2.8 Safety Compliance
1.2.9 Respirator Program
1.2.10 Industrial Hygienist

1.3 SUBMITTALS

1.3.1 Certificates of Compliance
1.3.2 Asbestos Plan
1.3.3 Testing Laboratory
1.3.4 Industrial Hygienist
1.3.5 Monitoring Results
1.3.6 Monitoring of Nonfriable Materials
1.3.7 Notification
1.3.8 Landfill
1.3.9 Pressure Differential Recordings Local Exhaust System
1.3.10 Training

2. EXECUTION

2.1 EQUIPMENT

2.1.1 Respirators
2.1.2 Special Clothing
2.1.3 Change Rooms

PART 1 - GENERAL

1.1 APPLICABLE PUBLICATIONS: The publications listed below form a part of this specification to the extent referenced. The publications are referred to in the text by the basic designation only.

1.1.1 Military Standard (Mil. Std.): MIL-STD-101B Color Code for Pipelines and for Compressed-Gas Cylinders

1.1.2 Federal Standard (Fed. Std.): Fed. Std. 595A & Notice 9 Colors

1.1.3 Code of Federal Regulations (CFR) Publications:

29 CFR 1926. 58 Asbestos, Tremolite, Anthophyllite, Actinolite
29 CFR 1910. 134 Respiratory Protection
29 CFR 1910. 145 Specifications for Accident Prevention Signs and Tags
40 CFR 61, Subpart A General Provisions

40 CFR 61, Subpart M National Emission Standard for Asbestos

1.1.4 American National Standard Institute (ANSI) Publications:
9.2-79 Fundamentals Governing the Design and Operation of Local Exhaust Systems
88.2-80 Practices for Respiratory Protection

1.2 REMOVAL AND DISPOSAL:

1.2.1 Description of Work: The work covered by this section includes the handling of friable materials containing asbestos which are encountered during removal and demolition operations and the incidental procedures and equipment required to protect workers and occupants of the building or area, or both, from contact with airborne asbestos fibers. The work also includes the disposal of the removed asbestos containing materials (ACM). The asbestos work includes the demolition and removal of (__________) located (____________). The asbestos control area for the removal of (____________) shall be considered (____________).

OR

1.2.1 Description of Work: The work covered by this section includes the handling of nonfriable materials containing asbestos which are encountered during removal and demolition operations and the incidental procedures and equipment required to protect workers and occupants of the building or area, or both, from contact with airborne asbestos fibers. The work also includes the disposal of the removed ACM. The asbestos work includes the demolition and removal of (____________) located (____________). Under normal conditions this material would not be considered hazardous; however, this material will release airborne concentrations of asbestos fibers during demolition and removal and therefore shall be handled in accordance with the removal and disposal procedures as specified herein. The asbestos control area for the removal of (____________) shall be considered (____________).

1.2.2 Definitions:

1.2.2.1 Amended Water: Water containing a wetting agent or surfactant.

1.2.2.2 Asbestos: The term asbestos includes chrysotile, amosite, crocidolite, tremolite, anthophyllite, and actinolite.

1.2.2.3 Asbestos Control Area: An area where asbestos removal operations are performed which is isolated by physical boundaries to prevent the spread of asbestos dust, fibers, or debris.

1.2.2.4 Asbestos Fibers: This expression refers to asbestos fibers having an aspect ratio of at least 3:1 and longer than 5 micrometers.

1.2.2.5 Area Monitoring: Sampling of asbestos fiber concentrations within the asbestos control area and outside the asbestos control area which is representative of the airborne concentrations of asbestos fibers in the theoretical breathing zone.

1.2.2.6 Asbestos Permissible Exposure Limit: 0.2 fibers (longer than 5 micrometers) per cubic centimeter expressed as an 8-hour time-weighted average.

1.2.2.7 Background: Airborne asbestos concentration in an area prior to any asbestos abatement activity.

1.2.2.8 Friable Asbestos Material: Material that contains more than 1% asbestos by weight and that can be crumbled, pulverized, or reduced to powder by hand pressure when dry.

1.2.2.9 HEPA Filter Equipment: High-efficiency particulate absolute filtered vacuuming equipment with a filter system capable of collecting and retaining asbestos fibers. Filters shall be of 99.97% efficiency for retaining fibers of 0.3 μm or larger.

1.2.2.10 Nonfriable Asbestos Material: Material that contains asbestos in which the fibers have been locked in by a bonding agent, coating, binder, or other material so that the asbestos is well bound and will not release fibers in excess of the asbestos control limit during any appropriate use, handling, storage, or transportation.

1.2.2.11 Personal Monitoring: Air sampling to determine asbestos fiber concentrations within the breathing zone of an employee.

1.2.2.12 Time-Weighted Average (TWA): The TWA is an 8-hour time-weighted average airborne concentration of fibers, longer than 5 μm, per cubic centimeter of air. Three samples are required to establish the 8-hour time-weighted average.

1.2.3 Title to Materials: All materials resulting from demolition work, except as specified otherwise, shall become the property of the Contractor and shall be disposed of as specified herein.

1.2.4 Protection of Existing Work to Remain: Perform demolition work without damage to or contamination of adjacent work. Where such work is damaged or contaminated, it shall be restored to its original condition.

1.2.5 Medical Requirements: 29 CFR 1926.58.

1.2.5.1 Medical Examinations: Before exposure to airborne asbestos fibers, provide workers with a comprehensive medical examination as required by 29 CFR 1926.58. This examination is not required if adequate records show the employee has been examined as required by 29 CFR 1926.58 requirements within the past year. The same medical examination shall be given on an annual basis to employees engaged in an occupation involving asbestos fibers and within 30 calendar days before or after the termination of employment in such occupation. Specifically identify X-ray films of asbestos workers to the consulting radiologist and mark medical record jackets with the word "ASBESTOS."

1.2.5.2 Medical Records: Maintain complete and accurate records of employees' medical examinations for a period of (40) (30) years after termination of employment and make records of the required medical examinations available for inspection and copying to: The Assistant Secretary of Labor for Occupational Safety and Health, The Director of The National Institute for Occupational Safety and Health (NIOSH), authorized representatives of either of them, and an employee's physician upon the request of the employee or former employee.

1.2.6 Training: Within three months prior to assignment to asbestos work, instruct each employee with regard to the hazards of asbestos, safety and health precautions, and the use and requirements for protective clothing and equipment including respirators. Fully cover engineering and other hazard control techniques and procedures.

1.2.7 Permits and Notifications: Secure necessary permits in conjunction with asbestos removal, hauling, and disposition and provide timely notification of such actions as may be required by federal, state, regional, and local authorities. Notify the Regional Office of the United States Environmental Protection Agency (USEPA) in accordance with 40 CFR 61 and provide copies of the notification to the contracting Officer and the State Environmental Regulatory Agency 20 days prior to commencement of the work.

1.2.8 Safety Compliance: In addition to detailed requirements of this specification, comply with laws, ordinances, rules, and regulations of federal, state, regional, and local authorities regarding handling, storing, transporting, and disposing of asbestos waste materials. Comply with Subparts A and M. Submit matters of interpretation of standards to the appropriate administrative agency for resolution before starting the work. Where the requirements of this specification and referenced documents vary, the most stringent requirement shall apply.

1.2.9 Respirator Program: Establish a respirator program as required by ANSI 88.2 and 29 CFR 1910.134.

1.2.10 Industrial Hygienist: Conduct monitoring and training under the direction of an Industrial Hygienist certified by the American Board of Industrial Hygiene.

1.3 SUBMITTALS: The following items shall be submitted to and approved by the Contracting Officer prior to commencing work involving asbestos materials.

1.3.1 Certification of Compliance: Show compliance with ANSI 9.2 for vacuums, ventilation equipment, and other equipment required to contain airborne asbestos fibers by providing manufacturers' certifications.

1.3.2 Asbestos Plan: Submit a detailed plan of work procedures to be used in the removal and demolition of materials containing asbestos. The plan shall be prepared, signed, and sealed, including certification number and date, by the Certified Industrial Hygienist. Such plan shall include location of asbestos control areas, change rooms, layout of change rooms, interface of trades involved in the construction, sequencing of asbestos related work, disposal plan, type of wetting agent and asbestos sealer to be used, air monitoring, and a detailed description of the method to be employed in order to control pollution. This plan must be approved prior to the start of any asbestos work. The Contractor and Certified Industrial Hygienist shall meet with the Contracting Officer prior to beginning work, to discuss in detail the asbestos plan, including work procedures and safety precautions.

1.3.3 Testing Laboratory: Submit the name, address, and telephone number of the testing laboratory selected for the monitoring, testing, and reporting of airborne concentrations of asbestos fibers along with certification that persons counting the samples have been judged proficient by successful participation within the last year in the National Institute for Occupational Safety and Health (NIOSH) Proficiency Analytical Testing (PAT) Program.

1.3.4 Industrial Hygienist: Submit the name, address, and telephone number of the Industrial Hygienist selected to prepare the asbestos plan, direct monitoring, and training, and proof that the Industrial Hygienist is certified by the American Board of Industrial Hygiene, including certification number and date.

1.3.5 Monitoring Results: Fiber counting shall be completed and results reviewed by the Industrial Hygienist within 16 hours. The Industrial Hygienist shall notify the Contractor and the Contracting Officer immediately of any exposures to asbestos fibers in excess of the acceptable limits. Submit monitoring results to the Contracting Officer within three working days, signed by the testing laboratory employee performing air monitoring, the employee who tested the sample, and the Certified Industrial Hygienist.

1.3.6 Monitoring of Nonfriable Materials: A copy of all monitoring reports of nonfriable asbestos, including a description of the work procedure at the time of air monitoring, shall be sent to Naval Facilities Engineering Command (Code 046) 200 Stovall St., Alexandria, VA 22332.

1.3.7 Notification: Notify the Contracting Officer three working days prior to the start of asbestos work. Notify the local fire department three days prior to removing fireproofing material from the building, and notify them when the new fireproofing material has been applied.

1.3.8 Landfill: Submit written evidence that the landfill for disposal is approved for asbestos disposal by the USEPA and state or local regulatory agency(s). Submit detailed delivery tickets, prepared, signed, and dated by an agent of the landfill, certifying the amount of asbestos materials delivered to the landfill, within three working days after delivery.

1.3.9 Pressure Differential Recordings Local Exhaust System: The local exhaust system shall be operated continuously 24 hours a day until the enclosure of the asbestos control area is removed. Pressure differential recordings for each work day shall be reviewed by the Industrial Hygienist and submitted to the Contracting Officer within 24 hours from the end of each work day. The Industrial Hygienist shall notify the Contractor and the Contracting Officer immediately of any variance in the pressure differential which could cause adjacent unsealed areas to have asbestos fiber concentrations in excess of 0.01 fibers per cubic centimeter.

1.3.10 Training: Submit certificates signed by each employee that the employee has received training in the proper handling of materials that contain asbestos; understands the health implications and risks involved, including the illnesses possible from exposure to airborne asbestos fibers; understands the use and limits of the respiratory equipment to be used; and understands the results of monitoring of airborne quantities of asbestos as related to health and respiratory equipment.

PART 2 - EXECUTION

2.1 EQUIPMENT: Make available to the Contracting Officer two complete sets of personal protective equipment as required herein for entry to the asbestos control area at all times for inspection of the asbestos control area.

2.1.1 Respirators: Select respirators from those approved by the Mine Safety and Health Administration (MSHA), Department of Labor, or the National Institute for Occupational Safety and Health (NIOSH), Department of Health and Human Services.

2.1.1.1 Respirators for Handling Asbestos: Provide personnel engaged in the removal and demolition of asbestos materials with Type C supplied-air respirators, continuous flow or pressure demand class.

2.1.1.2 Optional Respirators for Handling Asbestos: Use Type C continuous flow or pressure-demand, supplied-air respirators until the Contractor establishes the average airborne concentrations of asbestos the employees will confront. When the exposure limits are established, the respirators presented in 29 CFR 1926.58 that afford adequate protection at such upper concentrations of airborne asbestos may be used. If the Contractor decides to provide respirators other than a Type C continuous flow or pressure-demand, supplied-air respirator, the Contractor shall determine the exposure of each employee to airborne asbestos during each type of removal operation. Determine the 8-hour time-weighted average concentration of asbestos to which each of the employees is exposed during each type of removal operation.

2.1.2 Special Clothing:

2.1.2.1 Protective Clothing: Provide personnel exposed to airborne concentrations of asbestos fibers with fire-retardant, disposable, protective whole-body clothing, headcoverings, gloves, and foot coverings. Provide disposable plastic or rubber gloves to protect hands. Cloth gloves may be worn inside the plastic or rubber gloves for comfort, but shall not be used alone. Make sleeves secure at the wrists and make foot coverings secure at the ankles by the use of tape.

2.1.2.2 Work Clothing: Provide cloth work clothes for wear under the disposable protective coveralls and foot coverings.

2.1.3 Change Rooms: Provide a temporary unit with a separate decontamination locker room and a clean locker room for personnel required to wear whole-body protective clothing. Provide two separate lockers for each asbestos worker, one in each locker room. Keep street clothing and street shoes in the clean locker. Vacuum and remove asbestos-contaminated disposable protective clothing while still wearing respirators at the boundary of the asbestos work area and seal in impermeable bags or containers for disposal. Do not remove disposable protective clothing in the decontamination locker room. Remove cloth work clothing in the decontamination room. Tag and bag cloth work clothes for laundering and keep work shoes in the decontamination locker. Do not wear work clothing between home and work. Locate showers between the decontamination locker room and the clean locker room and require that all employees shower before changing into street clothes. Clean asbestos-contaminated work clothing in accordance with 29 CFR 1926.58. (Change rooms shall be physically attached or directly adjacent to the asbestos control area.)

2.1.4 Eye protection: Provide goggles to personnel engaged in asbestos operations when the use of a full-face respirator is not required.

2.1.5 Caution Signs and Labels: Provide (bilingual) caution signs at all approaches to asbestos control areas containing concentrations of airborne asbestos fibers. Locate signs at such a distance that personnel may read the sign and take the necessary protective steps required before entering the area. Provide labels and affix to all asbestos materials, scrap, waste, debris, and other products contaminated with asbestos.

2.1.5.1 Caution Signs: A vertical format conforming to 29 CFR 1926.58, minimum 20 by 14 in., displaying the following legend in the lower panel:

DANGER

ASBESTOS

CANCER AND LUNG DISEASE
HAZARD

AUTHORIZED PERSONNEL ONLY

RESPIRATORS AND PROTECTIVE
CLOTHING
ARE REQUIRED IN THIS AREA

Spacing between lines shall be at least equal to the height of the upper of any two lines.

2.1.5.2 Caution Labels: Provide labels of sufficient size to be clearly legible, displaying the following legend:

DANGER
CONTAINS ASBESTOS FIBERS
AVOID CREATING DUST
CANCER AND LUNG DISEASE
HAZARD

2.1.6 Tools and Local Exhaust System: Provide a local exhaust system in the asbestos control area. The local system shall be in accordance with ANSI Z 9.2. Equip exhaust with absolute (HEPA) filters. Local exhaust equipment must be sufficient to maintain a minimum pressure differential of minus 0.02 in. of water column relative to adjacent, unsealed areas. Provide continuous 24-hr per day monitoring of the pressure differential with an automatic recording instrument. In no case shall the building ventilation system be used as the local exhaust system for the asbestos control area. Filters on vacuums and exhaust equipment shall conform to ANSI 9.2.

2.2 WORK PROCEDURE: Perform asbestos related work in accordance with 29 CFR 1926.58 and as specified herein. Use (wet) (or) (dry) removal procedures. Personnel shall wear and utilize protective clothing and equipment as specified herein. Eating, smoking, or drinking shall not be permitted in the asbestos control area. Personnel of other trades not engaged in the removal and demolition of asbestos shall not be exposed at any time to provisions of this specification or complied with by the trade personnel. Shut down the building heating, ventilation, and air conditioning system (and provide temporary heating, ventilation, and air conditioning). (Disconnect electrical service when wet removal is performed and provide temporary electrical service.)

2.2.1 Furnishings: Furniture (______________) and equipment will be removed from the area of work by the Government before asbestos work begins.

OR

2.2.1 Furnishings: Furniture (______________) and equipment will remain in the building. Cover all furnishings with 6-mil plastic sheet or remove from the work area and store in a location on site approved by the Contracting Officer.

OR

2.2.1 Furnishings: If (furniture,) (books,) (equipment,) (carpeting,) (draperies,) (venetian blinds,) and (____________) in the work area (is) (are) contaminated with asbestos fibers, transfer these items to an area on site approved by the Contracting Officer set aside for cleaning (wet cleaning where possible) and then store until the room from which they came is declared clean and safe for entry. (Carpets, draperies, and other items which may not be susceptible to on-site wet cleaning methods shall be laundered in accordance with 29 CFR 1926.58.) At the conclusion of the asbestos removal work and cleanup operations, transfer all objects so removed and cleaned back to the area from which they came and reinstall them.

2.2.2 Masking and Sealing:

2.2.2.1 Asbestos Control Area Requirements: Seal openings in areas where the release of airborne asbestos fibers is expected. Establish an asbestos control area with the use of curtains, portable partitions, or other enclosures in order to prevent the escape of asbestos fibers from the contaminated asbestos control area. In all possible instances, control area development shall include protective covering of walls and ceilings with a continuous membrane of two layers of minimum 4-mil plastic sheet sealed with tape to prevent water or other damage. Provide two layers of 6-mil plastic sheet over floors and extend a minimum of 12 in. up walls. Seal all joints with tape. Provide a local exhaust system in the asbestos control area. Openings will be allowed in enclosures of asbestos control areas for the supply and exhaust of air for the local exhaust system. Replace filters as required to maintain the efficiency of the system.

2.2.2.2 Asbestos Control Area Requirements: The construction of an enclosed asbestos control area is impractical for the removal of (____________) located (____________). Establish designated limits for the asbestos work area. With the use of rope or other continuous barriers maintain all other requirements for asbestos control areas. Also, where an enclosure is not provided, conduct area monitoring of airborne asbestos fibers during the work shift at the designated limits downwind of the asbestos work area at such frequency as recommended by the Industrial Hygienist. If the quantity of airborne asbestos fibers monitored at the designated limits at any time exceeds background or 0.1 fibers/cm^3, whichever is less, evacuate personnel in adjacent areas or provide personnel with approved protective equipment. If adjacent areas are contaminated, clean the contaminated areas, monitor, and visually inspect the area as specified herein.

2.2.3 Asbestos Handling Procedures:

2.2.3.1 General Procedures: Sufficiently wet asbestos material with a fine spray of amended water during removal, cutting, or other handling so as to reduce the emission of airborne fibers. Remove material and immediately place in plastic disposal bags. Where unusual circumstances prohibit the use of plastic bags, submit an alternate proposal for containment of asbestos fibers to the Contracting Officer for approval. For example, in the case where both piping and insulation are to be removed, the Contractor may elect to wet the insulation and wrap the pipes and insulation in plastic and remove the pipe by sections.

2.2.3.2 Sealing and Removal of Asbestos-Contaminated Items Designated for Disposal: Remove contaminated architectural, mechanical, and electrical appurtenances such as venetian blinds, full-height partitions, carpeting, duct work, pipes and fittings, radiators, light fixtures, conduit, panels, and other contaminated items designated for removal by completely coating the items with an asbestos sealer at the demolition site before removing the items from the asbestos control area. Remaining asbestos residue shall not be of such size so as to allow dislodging by means other than vacuuming. These items need not be vacuumed. The asbestos sealer shall be tinted a contrasting color. It shall be spray-applied by airless method. Thoroughness of sealing operation shall be visually gauged by the extent of colored coating on exposed surfaces. Sealers shall be equal to the following products.

a. "Asbestite 2000," manufactured by Arpin Products, Inc. PO Box 262 Oak Hurst, NJ 07755, phone (201) 531-0674. (Application rate as recommended by manufacturer.)

b. "Wedbestos Sealer," available from Webco Products, Stinnes Western Chemical, 3270 East Washington Blvd., Los Angeles, CA 90023, phone (213) 269-0191. (Apply as recommended by manufacturer.)

c. "Dust-Set," manufactured by Mateson Chemical Corp., 1025 East Montgomery Ave., Philadelphia, PA 19125, phone (215) 423–3200. (Apply as recommended by manufacturer.)

2.2.3.3 Exposed Pipe Insulation Edges: Contain edges of asbestos insulation to remain that is exposed by a removal operation. Wet and cut the rough ends true and square with sharp tools and then encapsulate the edges with a 1/4-in.- thick layer of insulating cement troweled to a smooth hard finish. When cement is dry, lag the end with a layer of fiberglass cloth, overlapping the existing ends by 4 in. When insulating cement and cloth is an impractical method of sealing a raw edge of asbestos, take appropriate steps to seal the raw edges as approved by the Contracting Officer.

2.2.4 Identification of Asbestos-Free Insulation: Apply "ASBESTOS-FREE" markings to the exterior jacket of all nonasbestos insulated piping installed under this contract. Letter size shall be in accordance with MIL-STD-101. Apply such markings at maximum of 20-ft intervals. Indicate the limits of new "ASBESTOS-FREE" insulation with a 1-in.-wide band with attached arrow pointing in the direction of the label "ASBESTOS-FREE." Paint markings in orange, No. 12246 of Fed. Std. 595, and as specified in Section 09910, "Painting of Buildings (Field Painting)."

2.2.5 Monitoring: Monitoring of airborne concentrations of asbestos fibers shall be in accordance with 29 CFR 1926.58 and as specified herein.

2.2.5.1 Monitoring Prior to Asbestos Work: Provide area monitoring and establish the background one day prior to the masking and sealing operations for each demolition site.

2.2.5.2 Monitoring During Asbestos Work: Provide personal and area monitoring and establish the maximum TWA airborne concentration during the first active exposure to asbestos. Thereafter, provided the same type of work is being performed, provide area monitoring once every four hours during the work shift inside the asbestos control area, outside the entrance to the asbestos control area, and at the exhaust opening of the local exhaust system. If monitoring outside the asbestos control area shows airborne levels have exceeded background or 0.1 fibers/cm^3, whichever is less, stop all work, correct the condition(s) causing the increase, and notify the Contracting Officer immediately. (In areas where the construction of an asbestos control area is not required, after initial TWAs are established and provided the same type of work is being performed, provide monitoring at the designated limits of the asbestos work area at such frequency as recommended by the Industrial Hygienist.)

2.2.5.3 Monitoring After Final Cleanup: Provide area monitoring of asbestos fibers and establish the airborne asbestos concentration of less than 0.01 fibers/cm^3 after final cleanup but before removal of the enclosure of the asbestos control area. Provide area monitoring and establish the airborne asbestos concentration 2 days, 5 days (15 days, and 30 days) after the enclosure of the asbestos control area is removed (or after final cleanup when an enclosure is not required). The fiber counts from these samples shall be less than 0.01 fibers/cm^3 or not greater than the background, whichever is less. Should any of the final samplings indicate a higher value, the Contractor shall take appropriate actions to reclean the area and shall repeat the monitoring.

2.2.6 Site Inspection: While performing asbestos removal work, the Contractor shall be subject to on-site inspection by the Contracting Officer who may be assisted by safety or health personnel. If the work is found to be in violation of this specification, the Contracting Officer will issue a stop work order to be in effect immediately and until the violation is resolved. Standby time required to resolve the violation shall be at the Contractor's expense.

2.3 CLEANUP AND DISPOSAL:

2.3.1 Housekeeping: Essential parts of asbestos dust control are housekeeping and cleanup procedures. Maintain surfaces of the asbestos control area free of accumulations of asbestos fibers. Give meticulous attention to restricting the spread of dust and debris; keep waste from being distributed over the general area. Do not blow down the space with compressed air. When asbestos removal is complete, all asbestos debris is removed from the worksite, and final cleanup is completed, certify the area as safe before the signs are removed. After final cleanup and acceptable airborne concentrations are attained but before the HEPA unit is turned off and the containment removed, remove all filters on the building HVAC system and provide new filters. Dispose of filters as asbestos-contaminated materials. Reestablish HVAC, mechanical, and electrical systems in proper working order. The Contracting Officer will visually inspect the affected surfaces for residual asbestos material and accumulated dust and the Contractor shall reclean all areas showing dust or residual asbestos materials. If recleaning is required, monitor the asbestos airborne concentration after recleaning. Notify the Contracting Officer before unrestricted entry is permitted. The Government shall have the option to perform monitoring to certify the areas are safe before entry is permitted.

2.3.2 Disposal of Asbestos:

2.3.2.1 Procedure for Disposal: Collect asbestos waste, scrap, debris, bags, containers, equipment, and asbestos-contaminated clothing which may produce airborne concentrations of asbestos fibers and place in sealed impermeable bags. Affix a caution label to each bag. Dispose of waste asbestos material at an Environmental Protection Agency (EPA) or state-approved sanitary landfill off Government property. For temporary storage, store sealed impermeable bags in asbestos waste drums or skips. An area for interim storage of asbestos-waste-containing drums or skips will be assigned by the Contracting Officer or his authorized representative. Procedure for hauling and disposal shall comply with 40 CFR 61 (Subpart M), state, regional, and local standards. Sealed plastic bags may be dumped from drums into the burial site unless the bags have been broken or damaged. Damaged bags shall remain in the drum and the entire contaminated drum shall be buried. Uncontaminated drums may be recycled. Workers unloading the sealed drums shall wear appropriate respirators and personal protective equipment when handling asbestos materials at the disposal site.

GENERAL NOTES

1. Do not refer to this guide specification in the project specification. Use it as a manuscript to prepare the project specifications. Edit and modify this guide specification to meet project requirements. Where "as shown," "as indicated," "as detailed," or words of similar import are used, include all requirements so designated on the project drawings.

2. Do not include the following parts of this NFGS in the project specification:

- a. table of contents
- b. sketches
- c. general notes
- d. technical notes
- e. other supplemental information, if any, attached to this guide specification

As the first step in editing this guide specification for inclusion in a project specification, detach all parts listed above and, where applicable, use them in the editing process. If required in the construction contract, sketches and figures shall be placed on the project drawings. Where there are no project drawings, sketches and figures may be included as a part of the project specification if required.

3. Each capital letter in the right-hand margin of the text indicates that there is a technical note pertaining to that portion of the guide specification. Do not include these letters in the project specification. If this is a regionally tailored version of this NFGS, i.e. an EFD Regional Criteria Master, some technical notes and their designating letters may have been deleted.

4. Where numbers, symbols, words, phrases, clauses, sentences, or paragraphs in this guide specification are enclosed in parentheses, ( ), a choice or modification must be made; delete inapplicable portion(s). Where lines enclosed in brackets occur, insert

appropriate data. Delete inapplicable paragraphs and renumber subsequent paragraphs accordingly.

5. Project specification number, section number, and page numbers shall be centered at the bottom of each page of the section created from this guide specification.

EXAMPLE:
04-87-0000
02075-1

6. CAUTION: Coordination of this section with other sections of the project specification and with the drawings is mandatory. If materials or equipment are to be furnished under this section and installed under other sections or are indicated on the drawings, state that fact clearly for each type of material and item of equipment. Review the entire project specification and drawings to ensure that language is included to provide complete and operational systems and equipment.

7. Specifications shall not repeat information shown on the drawings. Specifications shall establish the quality of materials and workmanship, methods of installation, equipment functions, and testing required for the project. Drawings shall indicate dimensions of construction, relationship of materials, quantities, and location and capacity of equipment.

8. The project drawings shall clearly show location, extent, and form of asbestos materials to be removed or in contact with other removals or new work.

9. Suggestions for improvement of this specification will be welcomed. Complete the attached DD Form 1426 and mail the original to:

Commanding Officer
Northern Division, (Code 04AB)
Naval Facilities Engineering Command
Bldg. 77-Low, U.S. Naval Base
Philadelphia, PA 19112-5094

Mail a copy to:

COMMANDER
Naval Facilities Engineering Command
Code 04M2B
200 Stovall Street
Alexandria, VA 22332-2300

TECHNICAL NOTES

A. This guide specification covers the safety procedures and requirements for the removal and demolition of friable material containing asbestos. Nonfriable materials containing asbestos normally do not require special handling and disposal procedures unless such materials are sawn, pulverized, or handled in such a manner that will cause dust and airborne asbestos fibers to be released. If the project contains nonfriable asbestos that is considered to be hazardous due to material condition (broken down or excessively old and decayed) or demolition procedures to be used, then the nonfriable asbestos shall be specified to be removed in accordance with procedures established herein for friable asbestos. In such case, the wording in the second paragraph 1.2.1 should be used. On small asbestos removal operations an "enclosed" asbestos control area may not be required. The location of the area, type of material, and potential hazard must be reviewed and a judgment made by the designer. In a case where an enclosed area is not provided, many of the requirements in this specification should be deleted (see Notes E, O, and S).

NAVFAC policy is to eliminate the use of materials containing asbestos in cases where asbestos-free materials are available and suitable for the intended use; however, this is not intended to eliminate the use of asbestos materials completely since some items containing asbestos have no suitable substitute. The specifier should strive wherever possible to limit the use of asbestos in accordance with this policy.

When specifying the removal of asbestos materials, retain the following in paragraph 1.2 of Section 02050, "Demolition and Removal": "The demolition and removal of materials containing asbestos shall be in accordance with Section 02075, Removal and Disposal of Asbestos Materials."

B. Paragraph 1.1: The latest issue of applicable publications shall be used, but only after reviewing the latest issue to ensure that it will satisfy the minimum essential requirements of the project. If the latest issue of a referenced publication does not satisfy project requirements:

1. Use the issue shown; or
2. Select and refer to a document which does; or
3. Incorporate the pertinent requirements from the document into the project specification.

Delete those publications not referred to in the text of the section created from this guide specification.

C. Paragraph 1.2.1 (First Option): Use this paragraph for projects with demolition or removal of friable asbestos insulation. The limits of asbestos material in this category are asbestos insulation. The limits of asbestos removal must be indicated on the drawings or in the specification in sufficient detail for the Contractor to submit an accurate bid. Portions of the building where asbestos work will take place must be unoccupied during the removal operation. It is highly recommended in order to reduce exposure risk to occupants of the building that the entire building be unoccupied during asbestos removal operations. If portions of the building where asbestos removal is not taking place must remain occupied, additional requirements must be added for temporary heating/cooling and other utilities to the occupied portions of the building. The building heating/cooling system for example cannot be operated in the asbestos control area and due to wet removal procedures, electrical service to the asbestos control area may need to be shut off.

D. Paragraph 1.2.1: Specify the asbestos material to be removed in the first blank and the location of the material in the second blank. Example: "The asbestos work includes the demolition and removal of asbestos insulation located on existing steam piping indicated to be removed in the boiler room."

E. Paragraph 1.2.1: Specify the asbestos material to be removed in the first blank and identify the asbestos control area in the second blank. The asbestos control area is the area where asbestos operations are performed and is isolated by physical boundaries to prevent the spread of asbestos dust, fibers, or debris (see paragraph 2.2.2.1). Example: "The asbestos control area for the removal of asbestos insulation shall be considered to be the boiler room." If an "enclosed" asbestos control area is impractical and therefore not required, specify a distance from the asbestos work for the establishment of boundaries of the asbestos control area (see Note O).

F. Paragraph 1.2.1 (Second Option): Use this paragraph for projects where the demolition or removal of nonfriable asbestos is considered to be hazardous. The following materials are a few examples of nonfriable asbestos materials which are hazardous during removal:

a. Cement asbestos siding and roofing releases asbestos dust when sawn, broken, drilled, or sanded. Due to its brittleness, it is extremely difficult to remove cement asbestos without breakage and the release of asbestos dust.
b. Plaster materials with asbestos fibers, sprayed or troweled types.
c. Vinyl asbestos floor tile releases asbestos dust when scraping and grinding is required to remove tile and glue residue to level the existing subfloor.

Portions of the building where asbestos work will take place must be unoccupied during the removal operation. It is highly recommended in order to reduce exposure risk to occupants of the building that the entire building be unoccupied during asbestos removal operations. If portions of the building where asbestos removal is not taking place must remain occupied, additional requirements must be added for temporary heating/cooling and other utilities to the occupied portions of the building. The building heating/cooling system, for example, cannot be operated in the asbestos control area and due to wet removal procedures, electrical service to the asbestos control area may need to be shut off.

G. Paragraph 1.2.5.2: OSHA 29 CFR 1926.58 requires that medical records be retained at least 30 years. Some states require longer retention periods, the maximum being 40 years. Check with the state in which the project is located for the required retention time.

H. Paragraph 1.2.7: This paragraph may not be applicable for some overseas locations due to local government requirements. Verify the need for this requirement with the station safety officer.

I. Paragraph 1.3: Editing should include the following consideration:

1. Do not include the vertical bars in the final manuscript of the project specification. (Project Submittals Lists may be automatically extracted from project specifications prepared on NAVFAC-programmed word processors. Vertical bars indicate points at which automatically extracted entries will terminate.)

J. Paragraph 1.3.8: Use this paragraph for all projects except projects in overseas locations where local government regulations do not require a USEPA approved landfill.

K. Paragraphs 1.3.9 and 2.1.6: When an enclosed asbestos control area is not required, delete the requirements for the local exhaust system and pressure differential recording (see Note O).

L. Paragraph 2.1: Modify the number of sets of protective equipment as required, depending on the size of the asbestos removal project.

M. Paragraph 2.2: Use wet removal procedures in most cases. Wet removal is the preferred method and the least hazardous. Dry removal is an option where wet removal may damage adjacent areas. Dry removal as the only method of removal should only be specified if severe water damage to equipment, etc., would occur during wet removal. If dry removal alone is allowed, carefully edit the specification to remove all reference to amended water and wetting down procedures.

N. Paragraph 2.2.1: In most projects, the Government will remove furniture and equipment before the Contractor begins work. In this case the first paragraph should be used. The third paragraph should only be used when existing furnishings have been contaminated with asbestos fibers and the Contractor will be required to clean these items.

O. Paragraphs 2.2.2.1 and 2.2.2.2: When an "enclosed" asbestos control area is impractical, such as for the removal of cement asbestos roofing, or on a very small friable asbestos removal operation, use paragraph 2.2.2.2 and delete paragraph 2.2.2.1. If the project has both areas which can be enclosed and areas which cannot be enclosed, retain both paragraphs and identify the areas which must be enclosed and the areas which cannot be enclosed.

P. Paragraph 2.2.2.2: Specify the asbestos material to be removed in the first blank and identify the location of the area which cannot be enclosed in the second blank.

Q. Paragraph 2.2.3.2: Use this paragraph only when asbestos-contaminated items are also designated for removal and disposal.

R. Paragraphs 2.2.3.3 and 2.2.4: These paragraphs shall be included when new hot insulated piping is provided in a renovation project in which some existing asbestos insulation will remain in the building. This paragraph is not intended for new buildings or existing buildings without asbestos pipe insulation. This paragraph shall be coordinated with the Section 09910, "Painting of Buildings (Field Painting)."

Also, a reference to the marking requirements of this section should be noted in the section in which the insulation is specified.

S. Paragraphs 2.2.5.2 and 2.2.5.3: When an "enclosed" asbestos control area is not required, retain the portion in brackets.

T. Paragraph 2.3.2: Disposal procedures and sites for asbestos materials vary considerably with each location. Contact local station Public Works and the NAVFAC Division Hazardous Waste Manager or Industrial Hygienist for local procedures.

7

Work Practices for Removal Projects

This section provides guidelines, including recommended specifications and operating procedures, for use of negative-pressure systems in removing asbestos-containing materials (ACM) from buildings. A negative-pressure system is one in which static pressure in an enclosed work area is lower than that of the environment outside the containment barriers.

The pressure gradient is maintained by moving air from the work area to the environment outside the area via powered exhaust equipment at a rate that will support the desired air flow and pressure differential. Thus, the air moves into the work area through designated access spaces and any other barrier openings. Exhaust air is filtered by a high-efficiency particulate air (HEPA) filter to remove asbestos fibers.

The use of negative pressure during asbestos removal protects against large-scale release of fibers to the surrounding area in case of a breach in the containment barrier.

A negative-pressure system also can reduce the concentration of airborne asbestos in the work area by increasing the dilution ventilation rate (i.e., diluting contaminated air in the work area with uncontaminated air from outside) and exhausting contaminated air through HEPA filters. The circulation of fresh air through the work area reportedly also improves worker comfort, which may aid the removal process by increasing job productivity.

MATERIALS AND EQUIPMENT

The Portable, HEPA-Filtered, Powered Exhaust Unit

The exhaust unit establishes lower pressure inside than outside the enclosed work area during asbestos abatement. Basically, a unit consists of a cabinet with an opening at each end, one for air intake and one for exhaust. A fan and a series of filters are arranged inside the cabinet between the openings.

The fan draws contaminated air through the intake and filters and discharges clean air through the exhaust.

Portable exhaust units used for negative pressure systems in asbestos abatement projects should meet the following specifications.

Structural Specifications

The cabinet should be ruggedly constructed and made with durable materials to withstand damage from rough handling and transportation. The width of the cabinet should be less than 30 in. to fit through standard-size doorways. The cabinet must be appropriately sealed to prevent asbestos-containing dust from being emitted during use, transport, or maintenance. There should be easy access to all air filters from the intake end, and the filters must be easy to replace. The unit should be mounted on casters or wheels so it can be easily moved. It also should be accessible for easy cleaning.

Mechanical Specifications

Fans. The fan for each unit should be sized to draw a desired air flow through the filters in the unit at a specified static pressure drop. The unit should have an air-handling capacity of 1000 to

2000 ft^2/min (under "clean" filter conditions). The fan should be of centrifugal type.

For large-scale abatement projects, where the use of a larger capacity, specially designed exhaust system may be more practical than several smaller units, the fan should be appropriately sized according to the proper load capacity established for the application, i.e.,

$$\text{Total ft}^2/\text{min (load)} = \frac{\text{Volume of air in ft} \times \text{air changes/hr}}{60 \text{ min/hr}}$$

Smaller-capacity units (e.g., 1000 ft^2/min) equipped with appropriately sized fans and filters may be used to ventilate smaller work areas. The desired air flow could be achieved with several units.

Filters. The final filter must be the HEPA type. Each filter should have a standard nominal rating of at least 1100 ft^2/min with a maximum pressure drop of 1 in. H_2O clean resistance. The filter media (folded into closely pleated panels) must be completely sealed on all edges with a structurally rigid frame and cross-braced as required. The exact dimensions of the filter should correspond with the dimensions of the filter housing inside the cabinet or the dimensions of the filter-holding frame. The recommended standard size HEPA filter is 24 in. high × 24 in. wide × 11½ in. deep. The overall dimensions and squareness should be within ⅛ in.

A continuous rubber gasket must be located between the filter and the filter housing to form a tight seal. The gasket material should be ¼ in. thick and ¾ in. wide.

Each filter should be individually tested and certified by the manufacturer to have an efficiency of not less than 99.97% when challenged with 0.3-μm dioctylphthalate (DOP) particles. Testing should be in accordance with Military Standard Number 282 and Army Instruction Manual 136-300-175A. Each filter should bear a UL586 label to indicate ability to perform under specified conditions.

Each filter should be marked with: the name of the manufacturer, serial number, air flow rating, efficiency and resistance, and the direction of test air flow.

Prefilters, which protect the final filter by removing the larger particles, are recommended to prolong the operating life of the HEPA filter. Prefilters prevent the premature loading of the HEPA filter. They can also save energy and cost. One (minimum) or two (preferred) stages of prefiltration may be used. The first-stage prefilter should be a low-efficiency type (e.g., for particles 10 μm and larger). The second-stage (or intermediate) filter should have a medium efficiency (e.g., effective for particles down to 5 μm). Various types of filters and filter media for prefiltration applications are available from many manufacturers. Prefilters and intermediate filters should be installed either on or in the intake grid of the unit and held in place with special housings or clamps.

Instrumentation

Each unit should be equipped with a Magnehelic gauge or manometer to measure the pressure drop across the filters and indicate when filters have become loaded and need to be changed.

The static pressure across the filters (resistance) increases as they become loaded with dust, affecting the ability of the unit to move air at its rated capacity.

Electrical

General. The electrical system should have a remote fuse disconnect. The fan motor should be totally enclosed, fan-cooled, and the nonoverloading type. The unit must use a standard 115- V, single-phase, 60-cycle service. All electrical components must be approved by the National Electrical Manufacturers Association (NEMA) and Underwriter's Laboratories (UL).

Fans. The motor, the fan, the fan housing, and the cabinet should be grounded. The unit should have an electrical (or mechanical) lockout to prevent the fan from operating without a HEPA filter.

Instrumentation. An automatic shutdown system that would stop the fan in the event of a major rupture in the HEPA filter or blocked air discharge is recommended. Optional warning lights are recommended to indicate normal operation, too high a pressure drop across the filters (i.e., filter overloading), and too low a pressure drop (i.e., major rupture in HEPA filter or obstructed discharge). Other optional instruments include a timer and automatic shutoff and an elapsed time meter to show the total accumulated hours of operation.

Setup and Use of a Negative- Pressure System

Preparation of the Work Area

Determining the Ventilation Requirements for a Work Area. Experience with negative-pressure systems on asbestos abatement projects indicates a recommended rate of one air change every 15 min. The volume (in ft^2) of the work area is determined by multiplying the floor area by the ceiling height. The total air flow requirement (in ft^2/min) for the work area is determined by dividing this volume by the recommended air change rate (i.e., one air change every 15 min).

$$\text{Total ft}^2/\text{min} = \frac{\text{Volume of work area (in ft}^2)}{15 \text{ min}}$$

The number of units needed for the application is determined by dividing the total ft^2/min by the rated capacity of the exhaust unit.

$$\text{Number of units needed} = \frac{\text{Total ft}^2/\text{min}}{\text{Capacity of unit (in ft}^2)}$$

Location of Exhaust Units

The exhaust unit(s) should be located so that makeup air enters the work area primarily through the decontamination facility and traverses the work area as much as possible. This may be accomplished by positioning the exhaust unit(s) at a maximum distance from the worker access opening or other makeup air sources.

Wherever practical, work area exhaust units can be located on the floor in or near unused doorways or windows. The end of the unit or its exhaust duct should be placed through an opening in the plastic barrier or wall covering. The plastic around the unit or duct should then be sealed with tape.

Each unit must have temporary electrical power (115V AC). If necessary, three-wire extension cords can supply power to a unit. The cords must be in continuous lengths (without splice), in good condition, and should not be more than 100 ft long. They must not be fastened with staples, hung from nails, or suspended by wire. Extension cords should be suspended off the floor and out of workers' way to protect the cords from damage from traffic, sharp objects, and pinching.

Wherever possible, exhaust units should be vented to the outside of the building. This may involve the use of additional lengths of flexible or rigid duct connected to the air outlet and routed to the nearest outside opening. Windowpanes may have to be removed temporarily.

If exhaust air cannot be vented to the outside of the building or if cold temperatures necessitate measures to conserve heat and minimize cold air infiltration, filtered air that has been exhausted through the barrier may be recirculated into an adjacent area. However, this is not recommended.

Additional makeup air may be necessary to avoid creating too high a pressure differential, which could cause the plastic coverings and temporary barriers to "blow in." Additional makeup air also may be needed to move air most effectively through the work area. Supplemental makeup air inlets may be made by making openings in the plastic sheeting that allow air from outside the building into the work area. Auxiliary makeup air inlets should be as far as possible from the exhaust unit(s) (e.g., on an opposite wall), off the floor (preferably near the ceiling), and away from barriers that separate the work area from occupied clean areas. They should be resealed whenever the negative-pressure system is turned off after removal has started. Because the pressure differential (and ultimately the effectiveness of the system) is affected by the adequacy of makeup air, the number of auxiliary air inlets should be kept to a minimum to maintain negative pressure.

Use of the Negative-Pressure System

Testing the System. The negative-pressure system should be tested before any ACM is wetted or removed. After the work area has been prepared, the decontamination facility set up, and the exhaust unit(s) installed, the unit(s) should be started (one at a time). Observe the barriers and plastic sheeting. The plastic curtains of the decontamination facility should move slightly in toward the work area. The use of ventilation smoke tubes and a rubber bulb is another easy and inexpensive way to visually check system performance and direction of air flow through openings in the barrier. Another test is to use a Magnehelic gauge (or other instrument) to measure the static pressure differential across the barrier. The measuring device must be sensitive enough to detect a relatively low pressure drop. A Magnehelic gauge with a scale of 0 to 0.25 or 0.50 in. of H_20 and 0.005 or 0.01 in. graduations is generally adequate. The pressure drop across the barrier is measured from the outside by punching a small hole in the plastic barrier and inserting one end of a piece of rubber or Tygon tubing. The other end of the tubing is connected to the "low pressure" tap of the instrument. The "high pressure" tap must be open to the atmosphere. The pressure is read directly from the scale. After the test is completed, the hole in the barrier must be patched.

Use of System During Removal Operations. The exhaust units should be started just before beginning removal (i.e., before any ACM is disturbed). After removal has begun, the units should run continuously to maintain a constant negative pressure until decontamination of the work area is complete. The units should not be turned off at the end of the work shift or when removal operations temporarily stop.

Employees should start removing the asbestos material at a location farthest from the exhaust units and work toward them.

If an electric power failure occurs, removal must stop immediately and should not resume until power is restored and exhaust units are operating again.

Because airborne asbestos fibers are microscopic in size and tend to remain in suspension for a long time, the exhaust units must keep operating throughout the entire removal and decontamination processes. To ensure continuous operation, a spare unit should be available.

After asbestos removal equipment has been moved from the work area, the plastic sheeting has been cleaned, and all surfaces in the work area have been wet-cleaned, the exhaust units can be allowed to run for at least another 4 hr to remove airborne fibers that may have been generated during wet removal and cleanup and to purge the work area with clean makeup air. The units may be allowed to run for a longer time after decontamination, particularly if dry or only partially wetted asbestos material was encountered during removal.

Filter Replacement. All filters must be accessible from the work area or contaminated side of the barrier. Thus, personnel responsible for changing filters while the negative-pressure system is in use should wear approved respirators and other protective equipment. The operating life of a HEPA filter depends on the level of particulate contamination in the environment in which it is used. During use, filters will become loaded with dust, which increases resistance to air flow and diminishes the air-handling capacity of the unit. The difference in pressure drop across the filters between "clean" and "loaded" conditions is the usual indicator used to determine the replacement schedule for filters.

When pressure drop across the filters (as determined by the Magnehelic gauge or manometer on the unit) exceeds 1.0 in. of H_2O, the prefilter should be replaced first. The prefilter, which fan suction will generally hold in place on the intake grill, should be removed with the unit running by carefully rolling or folding in its sides. Any dust dislodged from the prefilter during removal will be collected on the intermediate filter. The used prefilter should be placed inside a plastic bag, sealed and labeled, and disposed of as asbestos waste. A new prefilter is then placed on the intake grill. Filters for prefiltration applications may be purchased as individual precut panels or in a roll of specified width that must be cut to size.

If the pressure drop still exceeds 1.0 in. of H_2O after the prefilter has been replaced, the intermediate filter must be replaced. With the unit operating, the prefilter should be removed, the intake grill or filter access opened, and their intermediate filter removed. Any dust dislodged from the intermediate filter during removal will be collected on the HEPA filter. The used intermediate filter should be placed in a sealable plastic bag (appropriately labeled) and disposed of as asbestos waste. A new replacement filter is then installed and the grill or access closed. Finally, the prefilter on the intake grill should be replaced.

The HEPA filter should be replaced if prefilter and/or intermediate filter replacement does not restore the pressure drop across the filters to its original clean resistance reading or if the HEPA filter becomes damaged. The exhaust unit must be shut off to replace the HEPA filter, which requires removing the prefilter first, then opening the intake grill or filter access, and finally removing the HEPA filter from the unit. Used HEPA filters should be placed in a sealable bag or container and a new filter properly rated and structurally identical to the original filter should then be installed. The intake grill and intermediate filter should be put back in place, the unit turned on, and the prefilter positioned on the intake grill. Whenever the HEPA filter is replaced, the prefilter and intermediate filter should also be replaced.

When several exhaust units are used to ventilate a work area, any air movement through an inactive unit during the HEPA filter replacement will be into the work area. Thus, the risk of asbestos fiber release to the outside environment is controlled.

Any filters used in the system may be replaced more frequently than the pressure drop across the filters indicates is necessary. Prefilters, for example, may be replaced two to four times a day or when accumulations of particulate matter become visible. Intermediate filters must be replaced once every day or so, and the HEPA filter may be replaced at the beginning of each new project. (Used HEPA filters must be disposed of as asbestos-containing waste.) Conditions in the work area dictate the frequency of filter changes. In a work area where fiber release is effectively controlled by thorough wetting and good work practices, fewer filter changes may be required than in work areas where the removal process is not well controlled. It should also be noted that the collection efficiency of a filter generally improves as particulate accumulates on it. Thus, filters can be used effectively until resistance (as a result of excessive particulate loading) diminishes the exhaust capacity of the unit.

Dismantling the System. When a final inspection and the results of final air tests indicate that the area has been decontaminated, all filters of the exhaust units should be removed and disposed of properly and the units shut off. The remaining barriers between contaminated and clean areas and all seals on openings into the work area and fixtures may be removed and disposed of as contaminated waste. A final check should be made to be sure that no dust or debris remain on surfaces as a result of dismantling operations.

WORK PRACTICES AND ENGINEERING CONTROLS FOR MAJOR ASBESTOS REMOVAL, RENOVATION, AND DEMOLITION OPERATIONS

This nonmandatory appendix, as published by OSHA, is designed to provide guidelines to assist employers in complying with the requirements of 29 CFR 1926.58. Specifically, this appendix describes the equipment, methods, and procedures that should be used in major asbestos removal projects conducted to abatc a recognized asbestos hazard or in preparation for building renovation or demolition. These projects require the construction of negative-pressure temporary enclosures to contain the asbestos material and to prevent the exposure of bystanders and other employees at the worksite.

The standard requires that "...whenever feasible, the employer shall establish negative-pressure enclosures before commencing asbestos removal, demolition or renovation operations." Employers should also be aware that when conducting asbestos removal projects they may be regulated under the National Emission Standards for Hazardous Air Pollutants (NESHAPS), 40 CFR Part 61, Subpart M, or EPA regulations under the Clear Water Act.

Construction of a negative-pressure enclosure is a simple but time-consuming process that requires careful preparation and execution; however, if the procedures below are followed contractors should be assured of achieving a temporary barricade that will protect employees and others outside the enclosure from exposure to asbestos and minimize to the extent possible the exposure of asbestos workers inside the barrier as well.

The equipment and materials required to construct these barriers are readily available and easily installed and used. In addition to an enclosure around the removal site the standard requires employers to provide hygiene facilities that ensure that their asbestos-contaminated employees do not leave the work site with asbestos on their persons or clothing; the construction of these facilities is also described below.

The steps in the process of preparing the asbestos removal site, building the enclosure, constructing hygiene facilities, removing the asbestos-containing material (ACM), and restoring the site include:

- planning the removal project
- procuring the necessary materials and equipment
- preparing the work area
- removing the ACM
- cleaning the work area
- disposing of the asbestos-containing waste

Planning the Removal Project

The planning of an asbestos removal project is critical to completing the project safely and cost-effectively. A written asbestos removal plan should be prepared describing the equipment and procedures that will be used throughout the project. The asbestos abatement plan will aid not only in executing the project but also in complying with the reporting requirements of the USEPA asbestos regulations (40 CFR 61. Subpart M) which call for specific information such as a description of control methods and control equipment to be used and the disposal sites the contractor proposes to use to dispose of the ACM.

The asbestos abatement plan should contain the following information:

- a physical description of the work area
- a description of the approximate amount of material to be removed
- a schedule for turning off and sealing existing ventilation systems
- personnel hygiene procedures
- labeling procedures
- a description of personal protective equipment and clothing to be worn by employees
- a description of the local exhaust ventilation systems to be used
- a description of work practices to be observed by employees
- a description of the methods to be used to remove the ACM
- the wetting agent to be used
- a description of the sealant to be used at the end of the project
- an air monitoring plan
- a description of the method to be used to transport waste material
- the location of the dump site

Materials and Equipment for Asbestos Removal

Although individual asbestos removal projects vary in terms of the equipment required to accomplish the removal of the material, some equipment and materials are common to most asbestos removal operations. Equipment and materials that should be available at the beginning of each project are: (1) rolls of polyethylene sheeting, (2) rolls of gray duct tape or clear plastic tape, (3) HEPA-filtered vacuum(s), (4) HEPA-filtered portable ventilation system(s), (5) a wetting agent, (6) an airless sprayer, (7) a portable shower unit, (8) appropriate respirators, (9) disposable coveralls, (10) signs and labels, (11) preprinted disposal bags, and (12) a manometer or pressure gauge.

Rolls of Polyethylene Plastic and Tape

Rolls of polyethylene plastic (6 mil in thickness) should be available to construct the asbestos removal enclosure and to seal windows, doors, ventilation systems, wall penetrations, and ceilings and floors in the work area. Gray duct tape or clear plastic tape should be used to seal the edges of the plastic and any holes in the plastic enclosure. Polyethylene plastic sheeting can be purchased in rolls up to 12-20 ft in width and up to 100 ft in length.

HEPA-Filtered Vacuum

A HEPA-filtered vacuum is essential for cleaning the work area after the asbestos has been removed. Such vacuums are designed to be used with a HEPA filter, which is capable of removing 99.97% of the asbestos particles from the air. Various sizes and capacities of HEPA vacuums are available. One manufacturer, Nilfisk of America, Inc., produces three models that range in capacity from 5.25 gal to 17 gal.

Exhaust Air Filtration System

A portable ventilation system is necessary to create a negative pressure within the asbestos removal enclosure. Such units are equipped with a HEPA filter and are designed to exhaust and clean the air inside the enclosure before exhausting it to the outside of the enclosure.

Systems are available from several manufacturers. One supplier, Micro-Trap Inc., has two ventilation units that range in capacity from 600 ft^2/min to 1700 ft^2/min. According to the manufacturer's literature, Micro-Trap units filter particles of 0.3 μm in size with an efficiency of 99.99%. The number and capacity of units required to ventilate an enclosure depend on the size of the area to be ventilated.

Wetting Agents

Wetting agents (surfactants) are added to water (which is then called amended water) and used to soak ACM; amended water penetrates more effectively than plain water and permits more thorough soaking of the ACM.

Wetting the ACM reduces the number of fibers that will break free and become airborne when the ACM is handled or otherwise disturbed. Asbestos-containing materials should be thoroughly soaked before removal is attempted; the dislodged material should feel spongy to the touch. Wetting agents are generally prepared by mixing 1–3 oz of wetting agent to 5 gal of water.

One type of asbestos, amosite, is relatively resistant to soaking, either with plain or amended water. The work practices of choice when working with amosite-containing material are to soak the material as much as possible and then to bag it for disposal immediately after removal, so that the material has no time to dry and be ground into smaller particles that are more likely to liberate airborne asbestos.

In a very limited number of situations it may not be possible to wet the ACM before removing it. Examples of such rare situations are: (1) removal of asbestos material from a "live" electrical box that was oversprayed with the material when the rest of the area was sprayed with asbestos-containing coating, and (2) removing asbestos-containing insulation from a live steam pipe. In both of these situations the preferred approach would be to turn off the electricity or steam respectively, to permit wet removal methods to be used. However, where removal work must be performed during working hours, i.e., when normal operations cannot be disrupted, the ACM must be removed dry. Immediate bagging is then the only method of minimizing the amount of airborne asbestos generated.

Airless Sprayer

Airless sprayers are used to apply amended water to ACM. Airless sprayers allow the amended water to be applied in a fine spray that minimizes the release of asbestos fibers by reducing the impact of the spray on the material to be removed. Airless sprayers are inexpensive and readily available.

Portable Shower

Unless the site has available a permanent shower facility that is contiguous to the removal area, a portable shower system is necessary to permit employees to clean themselves after exposure to asbestos and to remove any asbestos contamination from their hair and bodies. Taking a shower prevents employees from leaving the work area with asbestos on their clothes and thus prevents the spread of asbestos contamination to areas outside the asbestos removal area. This measure also protects members of the families of asbestos workers from possible exposure to asbestos. Showers should be supplied with warm water and a drain. A shower water filtration system to filter asbestos fibers from the shower water is recommended. Portable shower units are readily available, inexpensive, and easy to install and transport.

Respirators

Employees involved in asbestos removal projects should be provided with appropriate NIOSH-approved respirators. Selection of the appropriate respirator should be based on the concentration of asbestos fibers in the work area. If the concentration of asbestos fibers is unknown, employees should be provided with respirators that will provide protection against the highest concentration of asbestos fibers that can reasonably be expected to exist in the work area. For most work within an enclosure, employees should wear half-mask dual-filter cartridge respirators. Disposable face mask respirators (single-use) should not be used to protect employees from exposure to asbestos fibers.

Disposable Coveralls

Employees involved in asbestos removal operations should be provided with disposable impervious coveralls that are equipped with head and foot covers. Such coveralls are typically made of Tyvek. The coverall has a zipper front and elastic wrists and ankles.

Signs and Labels

Before work begins, a supply of signs to demarcate the entrance to the work area should be obtained. Signs are available that have the wording required by the final OSHA standard. The required labels are also commercially available as press-on labels and preprinted on the 6-mil polyethylene plastic bags used to dispose of asbestos-containing waste material.

Preparing the Work Area

Preparation for the construction of negative-pressure enclosures should begin with the removal of all movable objects from the work area, e.g., desks, chairs, rugs, and light fixtures, to ensure that these objects do not become contaminated with asbestos. When movable objects are contaminated or are suspected of being contaminated they should be vacuumed with a HEPA vacuum and cleaned with amended water unless they are made of material that will be damaged by the wetting agent; wiping with plain water is recommended in those cases where amended water will damage the object. Before the asbestos removal work begins, objects that cannot be removed from the work area should be covered with a 6-mil-thick polyethylene plastic sheeting that is securely taped with duct tape or plastic tape to achieve an airtight seal around the object.

Constructing the Enclosure

When all objects have either been removed from the work area or covered with plastic, all penetrations of the floor, walls and ceiling should be sealed with 6-mil polyethylene plastic and tape to prevent airborne asbestos from escaping into areas outside the work area or from lodging in cracks around the penetrations. Penetrations that require sealing are typically found around electrical conduits, telephone wires, and water supply and drain pipes. A single entrance to be used for access and egress to the work area should be selected, and all other doors and windows should be sealed with tape or be covered with 6-mil polyethylene plastic sheeting and securely taped. Covering windows and unnecessary doors with a layer of polyethylene before covering the walls provides a second layer of protection and saves time in installation because it reduces the number of edges that must be cut and taped. All other surfaces such as support columns, ledges, pipes, and other surfaces should also be covered with polyethylene plastic sheeting and taped before the walls themselves are completely covered with sheeting.

Next a thin layer of spray adhesive should be sprayed along the top of all walls surrounding the enclosed work area, close to the wall-ceiling interface, and a layer of polyethylene plastic sheeting should be stuck to this adhesive and taped. The entire inside surfaces of all wall areas are covered in this manner. The sheeting over the walls is extended across the floor area until it meets in the center of the area where it is taped to form a single layer of material enclosing the entire room, except for the ceiling. A final layer of plastic sheeting is then laid across the plastic-covered floor area and up the walls to a level of 2 ft or so; this layer provides a second protective layer of plastic sheeting over the floor which can then be removed and disposed of easily after the ACM that has dropped to the floor has been bagged and removed.

Building Hygiene Facilities

The standard mandates that employers involved in asbestos removal, demolition, or renovation operations provide their employees with hygiene facilities to be used to decontaminate asbestos exposed workers, equipment, and clothing before these leave the work area. These decontamination facilities consist of (1) a clean change room, (2) a shower, and (3) an equipment room.

The clean change room is an area in which employees remove their street clothes and don their respirators and disposable protective clothing. The clean room should have hooks on the wall or be equipped with lockers for the storage of workers' clothing and personal articles. Extra disposable coveralls and towels can also be stored in the clean change room.

The shower should be contiguous with both the clean and dirty change room and should be used by all workers leaving the work area. The shower should also be used to clean asbestos- contaminated equipment and materials such as the outsides of asbestos waste bags and hand tools used in the removal process.

The equipment room (also called the dirty change room) is the area where workers remove their protective coveralls and where equipment that is to be used in the work area can be stored. The equipment room should be lined with 6-mil thick polyethylene plastic sheeting in the same way as was done in the work area enclosure: two layers of 6-mil polyethylene plastic sheeting that are not taped together, form a double flap or barrier between the equipment room and the work area, and between the shower and the clean change room.

When feasible the clean change room, shower, and equipment room should be contiguous to the negative-pressure enclosure surrounding the removal area. In the majority of cases, hygiene facilities can be built contiguous to the negative-pressure enclosure. In some cases, however, hygiene facilities may have to be located on another floor of the building where removal of ACM is taking place. In these instances, the hygiene facilities can in effect be made to be contiguous to the work area by constructing a polyethylene plastic "tunnel" from the work area to the hygiene facilities. Such a tunnel can be made even in cases where the hygiene facilities are located several floors above or below the work area; the tunnel begins with a double-flap door at the enclosure, extends through the exit from the floor, continues down the necessary number of flights of stairs and goes through a double-flap entrance to the hygiene facilities which have been prepared as described above. The tunnel is constructed of 2-in. by 4-in. lumber or aluminum struts and covered with 6-mil thick polyethylene plastic sheeting.

In the rare instances when there is not enough space to permit any hygiene facilities to be built at the work site, employees should be directed to change into a clean disposable worksuit immediately after exiting the enclosure (without removing their respirators) and to proceed immediately to the shower. Alternatively, employees could be directed to vacuum their disposable coveralls with a HEPA-filtered vacuum before proceeding to a shower located a distance from the enclosure.

The clean room, shower, and equipment room must be sealed completely to ensure that the sole source of air flow through these areas originates from uncontaminated areas outside the asbestos removal, demolition, or renovation enclosure. The shower must be drained properly after each use to ensure that contaminated water is not released to uncontaminated areas. If waste water is inadvertently released, it should be cleaned up as soon as possible to prevent any asbestos in the water from drying and becoming airborne in areas outside the work area.

Establishing Negative Pressure Within the Enclosure

After construction of the enclosure is completed, a ventilation system(s) should be installed to create a negative pressure within the enclosure with respect to the area outside the enclosure. Such ventilation systems must be equipped with HEPA filters to prevent the release of asbestos fibers to the environment outside the enclosure and should be operated 24 hr per day during the entire project until the final cleanup is completed and the results of final air samples are received from the laboratory. A sufficient amount of air should be exhausted to create a pressure of -0.02 in. of water within the enclosure with respect to the area outside the enclosure.

These ventilation systems should exhaust the HEPA-filtered clean air outside the building in which the asbestos removal, demolition, or renovation is taking place.

If access to the outside is not available, the ventilation system can exhaust the HEPA-filtered asbestos-free air to an area within the building that is as far away as possible from the enclosure. Care should be taken to ensure that the clean air is released either to an asbestos-free area or in such a way as not to disturb any ACM.

A manometer or pressure gauge for measuring the negative pressure within the enclosure should be installed and should be monitored frequently throughout all work shifts during which asbestos removal, demolition, or renovation takes place. Several types of manometers and pressure gauges are available for this purpose.

All asbestos removal, renovation, and demolition operations should have a program for monitoring the concentration of airborne asbestos and employee exposures to asbestos. Area samples should be collected inside the enclosure (approximately four samples for 5000 ft^2 of enclosure area). At least two samples should be collected outside the work area, one at the entrance to the clean change room and one at the exhaust of the portable ventilation system. In addition, several breathing zone samples should be collected from those workers who can reasonably be expected to have the highest potential exposure to asbestos.

Removing Asbestos Materials

Employers involved in asbestos removal, demolition, or renovation operations designate a competent person to:

- set up the enclosure
- ensure the integrity of the enclosure
- control entry to and exit from the enclosure
- supervise all employee exposure monitoring required by this section
- ensure the use of protective clothing and equipment
- ensure that employees are trained in the use of engineering controls, work practices, and personal protective equipment
- ensure the use of hygiene facilities and the observance of proper decontamination procedures
- ensure that engineering controls are functioning properly

The competent person will generally be a Certified Industrial Hygienist, an industrial hygienist with training and experience in the handling of asbestos, or a person who has such training and experience as a result of on-the-job training and experience.

Ensuring the integrity of the enclosure is accomplished by inspecting the enclosure before asbestos removal work begins and prior to each work shift throughout the entire period work is being conducted in the enclosure. The inspection should be conducted by locating all areas where air might escape from the enclosure; this is best accomplished by running a hand over all seams in the plastic enclosure to ensure that no seams are ripped and the tape is securely in place. The competent person should also ensure that only authorized personnel enter the enclosure and that all employees and other personnel who enter the enclosure have the proper protective clothing and equipment. He or she should also ensure that all employees and other personnel who enter the enclosure use the hygiene facilities and observe the proper decontamination procedures.

Proper work practices are necessary during asbestos removal, demolition, and renovation to ensure that the concentration of asbestos fibers inside the enclosure remains as low as possible. One of the most important work practices is to wet the ACM before it is disturbed. After the ACM is thoroughly wetted, it should be removed by scraping (as in the case of sprayed-on or troweled-on ceiling material) or removed by cutting the metal bands or wire mesh that support the ACM on boilers or pipes. Any residue that remains on the surface of the object from which asbestos is being removed should be wire-brushed and wet-wiped.

Bagging asbestos waste material promptly after its removal is another work practice control that is effective in reducing the airborne concentration of asbestos within the enclosure. Whenever possible the asbestos should be removed and placed directly into bags for disposal rather than dropping the material to the floor and picking up all of the material when the removal is complete. If a significant amount of time elapses between the time the material is removed and the time it is bagged, the asbestos material is likely to dry out and generate asbestos-laden dust when it is disturbed by people working within the enclosure. Any asbestos-contaminated supplies and equipment that cannot be decontaminated should be disposed of in prelabeled bags; items in this category include plastic sheeting, disposable work clothing, respirator cartridges, and contaminated wash water.

A checklist is one of the most effective methods of ensuring adequate surveillance of the integrity of the asbestos removal enclosure.

Filling out the checklist at the beginning of each shift in which asbestos removal is being performed will serve to document that all the necessary precautions will be taken during the asbestos removal work. The checklist contains entries for ensuring that:

- the work area enclosure is complete
- the negative-pressure system is in operation
- necessary signs and labels are used
- appropriate work practices are used
- necessary protective clothing and equipment are used
- appropriate decontamination procedures are being followed

SECTION 1926.58—WORK PRACTICES AND ENGINEERING CONTROLS FOR SMALL-SCALE, SHORT-DURATION ASBESTOS RENOVATION AND MAINTENANCE ACTIVITIES[14]

Definition of Small-Scale, Short-Duration Activities

For the purposes of this book, small-scale, short-duration renovation and maintenance activities are tasks such as but not limited to:

- removal of asbestos-containing insulation on pipes
- removal of small quantities of asbestos-containing insulation on beams or above ceilings
- replacement of an asbestos-containing gasket on a valve
- installation or removal of a small section of drywall
- installation of electrical conduits through or proximate to ACM
- wet methods
- removal methods

    use of glove bags
    removal of entire asbestos-insulated pipes or structures
    use of mini-enclosures

enclosure of asbestos materials
maintenance programs

Preparation of the Area Before Renovation or Maintenance Activities

The first step in preparing to perform a small-scale, short-duration asbestos renovation or maintenance task, regardless of the abatement method that will be used, is the removal from the work area of all objects that are movable to protect them from asbestos contamination. Objects that cannot be removed must be covered completely with 6-mil thick polyethylene plastic sheeting before the task begins. If objects have already been contaminated they should be thoroughly cleaned with a high-efficiency particulate air (HEPA)-filtered vacuum or be wet-wiped before they are removed from the work area or completely encased in the plastic.

Wet Methods

Whenever feasible and regardless of the abatement methods to be used (e.g., removal, enclosure, use of glove bags) wet methods must be used during small-scale, short-duration maintenance and renovation activities that involve disturbing ACM. Handling asbestos materials wet is one of the most reliable methods of ensuring that asbestos fibers do not become airborne.

Wet methods can be used in the great majority of workplace situations. Only in cases where asbestos work must be performed on live electrical equipment, on live steam lines, or in other areas where water will seriously damage materials or equipment, may dry removal be performed. Amended water or another wetting agent should be applied by means of an airless sprayer to minimize the extent to which the ACM is disturbed.

Asbestos-containing materials should be wetted from the initiation of the maintenance or renovation operation, and wetting agents should be used continually throughout the work period to ensure that any dry ACM exposed in the course of the work is wet and remains wet until final disposal.

Removal of Small Amounts of ACM

There are several methods that can be used to remove small amounts of ACM during small-scale, short-duration renovation or maintenance tasks. These include the use of glove bags, the removal of an entire asbestos-covered pipe or structure, and the construction of mini-enclosures. The procedures that employers must use for each of these operations if they wish to avail themselves of the final rule's exemptions are described in the following sections.

Glove Bag Installation

Glove bags are approximately 40-in.-wide × 64-in.-long bags fitted with arms through which the work can be performed. When properly installed and used they permit workers to remain completely isolated from the asbestos material being removed or replaced inside the bag. Glove bags can thus provide a flexible, easily installed, and quickly dismantled, temporary small work area enclosure that is ideal for small-scale asbestos renovation and maintenance jobs.

These bags are single-use control devices that are disposed of at the end of each job. The bags are made of transparent 6-mil thick polyethylene plastic with arms of Tyvek material (the same material used to make the disposable protective suits used in major asbestos removal, renovation, and demolition operations and in protective gloves). Glove bags are readily available from safety supply stores or specialty asbestos removal supply houses. Glove bags come prelabeled with the asbestos warning label prescribed by OSHA and EPA for bags used to dispose of asbestos waste.

Glove Bag Equipment and Supplies

Supplies and materials that are necessary to use glove bags effectively include:

- tape to seal the glove bag to the area from which asbestos is to be removed

- amended water or other wetting agents
- an airless sprayer for the application of the wetting agent
- bridging encapsulant (a paste-like substance for coating asbestos) to seal the rough edges of any ACM that remains within the glove bag at the points of attachment after the rest of the asbestos has been removed
- tools such as razor knives, nips, and wire brushes (or other tools suitable for cutting wire, etc.)
- an HEPA filter-equipped vacuum for evacuating the glove bag (to minimize the release of asbestos fibers) during removal of the bag from the work area and for cleaning any material that may have escaped during the installation of the glove bag
- HEPA-equipped dust cartridge respirators for use by the employees involved in the removal of asbestos with the glove bag

A checklist for asbestos removal, renovation, and demolition is shown in Figure 1.

Cleaning the Work Area

After all of the ACM is removed and bagged, the entire work area should be cleaned until it is free of all visible asbestos dust. All surfaces from which asbestos has been removed should be cleaned by wire-brushing the surfaces, HEPA-vacuuming these surfaces, and wiping them with amended water. The inside of the plastic enclosure should be vacuumed with a HEPA vacuum and wet-wiped until there is no visible dust in the enclosure. Particular attention should be given to small horizontal surfaces such as pipes, electrical conduits, lights, and support tracks for drop ceilings. All such surfaces should be free of visible dust before the final air samples are collected.

Additional sampling should be conducted inside the enclosure after the cleanup of the work area has been completed. Approximately four area samples should be collected for each 5000 ft^2 of

Figure 1. Asbestos Removal, Renovation, and Demolition Checklist[15]

Date: ____________________ Location: ________________________________

Supervisor: ____________________________ Project # ____________________

Work area (ft^2) ________________

		YES	NO
I.	Work site barrier		
	Floor covered	____	____
	Walls covered	____	____
	Area ventilation off	____	____
	All edges sealed	____	____
	Penetrations sealed	____	____
	Entry curtains	____	____
II.	Negative Air Pressure		
	HEPA Vac ______________		
	Ventilation system ______________		
	Constant operation	____	____
	Negative pressure achieved	____	____
III.	Signs		
	Work area entrance	____	____
	Bags labeled	____	____
IV.	Work Practices	____	____
	Removed material promptly bagged	____	____
	Material worked wet	____	____
	HEPA vacuum used	____	____
	No smoking	____	____
	No eating, drinking	____	____
	Work area cleaned after completion	____	____
	Personnel decontaminated each	____	____
	departure	____	____
V.	Protective Equipment		
	Disposable clothing used one time	____	____
	Proper NIOSH-approved respirators	____	____

Figure 1, cont.

VI.	Showers		
	On site	_____	_____
	Functioning	_____	_____
	Soap and towels	_____	_____
	Used by all personnel	_____	_____

enclosure area. The enclosure should not be dismantled until the final samples show asbestos concentrations of less than the final standard's action level. EPA recommends that a clearance level of 0.01 ft/cm^3 be achieved before cleanup is considered complete.

A clearance checklist is an effective method of ensuring that all surfaces are adequately cleaned and the enclosure is ready to be dismantled. Figure 2 shows such a checklist.

Preparing the Work Area

Airborne fibers which are generated by disturbance of ACM may remain suspended in the air for long periods of time because of their small size and aerodynamic properties. These airborne asbestos fibers can migrate via air currents to other parts of the building.

Proper preparation of the work area before an asbestos abatement project begins serves the primary purpose of containing fibers which are released within the work area. Good preparation techniques serve to protect interior finishes such as hardwood floors or carpets from water damage and reduce cleanup effort. General safety issues are also a major consideration in work area preparation.

Each project has unique requirements for effective preparation. For instance, the sequence of steps would probably be different for preparing a boiler room than for preparing an area with asbestos material above a suspended ceiling. The following are general guidelines which can be modified to address specific problems encountered on an asbestos abatement project.

Figure 2. Final Inspection of Asbestos Removal, Renovation, and Demolition Projects

---

Date: ____________________

Project: ____________________

Location: ____________________

Building: ____________________

CHECKLIST:

Residual dust on:		Yes	No
a.	Floor	____	____
b.	Horizontal surfaces	____	____
c.	Pipes	____	____
d.	Ventilation equipment	____	____
e.	Horizontal surfaces	____	____
f.	Pipes	____	____
g.	Ducts	____	____
h.	Register	____	____
i.	Lights	____	____

FIELD NOTES:

Record any problems here.

____

____

____

____

____

____

____

____

____

FINAL AIR SAMPLE RESULTS: ____________________

____

____

____

STEP 1 — Conduct Walk-Through Survey of the Work Area

The contractor, building owner, and architect should make a walk-through survey to inventory and photograph any existing damages.

STEP 2 — Post Warning Signs

Warning signs should be placed at each entrance to the work area. Reusable metal signs or disposable cardboard signs are available. Signs should inform the reader that breathing asbestos dust may cause serious bodily harm. See the Occupational Safety and Health Administration asbestos standard for sign specification (Chapter 6). These signs are available from most safety supply houses and asbestos abatement contractor suppliers.

STEP 3 — Shut Down the Heating, Ventilating, and Air Conditioning (HVAC) System

The HVAC system supplying the work area should be shut down and isolated to prevent entrainment of asbestos dust throughout the building. To avoid inadvertent activation of the HVAC system while removal operations are in progress, the control panel should be tagged (advising personnel not to activate) and locked.

All vents and air ducts inside the work area should be covered and sealed with two layers of 6-mil polyethylene and duct tape. The first layer of polyethylene should be left in place until the area has passed final visual inspection and clearance air monitoring.

HVAC filters which may be contaminated with asbestos dust should be removed and disposed of in the same manner as the other ACM.

STEP 4 — Clean and Remove Furniture and Nonstationary Items from the Work Area

Workers wearing half-mask high-efficiency filter cartridge respirators and disposable clothing should remove all nonstationary items that can feasibly be taken out of the work area. This prevents further contamination of the items and facilitates the removal process. Before the items are stored outside the work area, they should be cleaned with a HEPA-filtered vacuum and/or wet-wiped to remove any asbestos-containing dust. Drapes should be removed for dry cleaning or disposal. Carpet should be disposed of as asbestos-containing waste.

STEP 5 — Seal Stationary Items with Polyethylene

Items not being removed from the work area, such as large pieces of machinery, blackboards, pencil sharpeners, water fountains, toilets, etc., should be wet-wiped or HEPA-vacuumed and wrapped in place with 6-mil polyethylene and sealed with duct tape.

Water fountains should be disconnected, covered with two layers of polyethylene, and labeled nonoperational to discourage anyone from cutting through the polyethylene to get a drink.

Electrical outlets should be shut down, if possible, and sealed with tape or covered with polyethylene and then taped.

STEP 6 — Tape and Seal Windows with Polyethylene

The edges of all the windows should be sealed with 3 in.-wide high quality duct tape. After the edges have been taped, the windows should be covered and sealed with 6-mil polyethylene and duct tape.

STEP 7 — Cover the Floor with Polyethylene

Polyethylene sheets (6 mil) should be used to cover the floor in the work area. Several sheets may be seamed together with spray adhesive and duct tape. Blue or red carpenter's chalk placed beneath the seam line will darken in color if water leaks through. Any leaks which occur should be promptly cleaned up. The polyethylene floor sheets should be cut and peeled back to allow access to the wet area. After mopping up the water and any contamination that leaked through, the area should be wet-wiped with clean rags. The peeled-back sheets are put back in place and sealed with duct tape after the area dries. An additional "patch" sheet can be placed over this area and sealed with tape to provide extra protection.

After joining the sheets of polyethylene together, the floor covering should be cut to the proper dimensions, allowing the polyethylene to extend 24 in. up the wall all the way around the room. The polyethylene should be flush with the walls at each corner to prevent damage by foot traffic.

When the first layer of polyethylene has been secured in place, a second layer should be installed with the seams of the first and second layers offset. The second layer of polyethylene should extend a few inches above the first layer on the wall and be secured with 3-in. duct tape.

When covering stairs, ramps, or other potential slippery spots with polyethylene, care must be taken to provide traction for foot traffic. Wet polyethylene is very slippery and can create serious tripping hazards. To provide better footing, masking tape or thin wood strips can be placed on top of the polyethylene to provide rough surfaces in these areas.

STEP 8 — Cover the Walls with Polyethylene

After the floors and stationary objects have been covered with polyethylene, one or two layers of 4-mil polyethylene are used to cover the walls. The lighter weight 4-mil is easier to hang and keep in place than the heavier 6-mil.

The sheets of 4-mil polyethylene should hang from the top of the wall a few inches below the asbestos material and should be long enough to overlap the floor sheets by 24 in. The vertical sheets should be overlapped and seams sealed with adhesive duct tape. The sheets should be hung using a combination of nails and furring strips (small wood blocks), or adhesive and staples, and sealed with 4-in. duct tape. Duct tape alone will not support the weight of the polyethylene after exposure to the high humidity which often occurs inside the work area. Nails may cause some minor damage to the interior finish; however, it is usually more time-efficient to touch up the nail holes than to repeatedly repair fallen barriers.

STEP 9 — Locate and Secure the Electrical System to Prevent Shock Hazards

Amended water is typically used to saturate asbestos-containing sprayed-on material prior to removal. This creates a humid environment with damp to very wet floors. The electrical supply to the work area should be de-energized and locked out before removal operations begin to eliminate the potential for a shock hazard. Before removal begins:

- identify and de-energize electrical circuits in the work area
- lock the breaker box after the system has been shut down and place a warning tag on the box
- make provisions for supplying the work area with electricity from outside the work area which is equipped with a ground-fault-interrupt system
- if the electrical supply cannot be disconnected, energized parts must be insulated or guarded from employee contact and any other conductive object

STEP 10 — Removing or Covering Light Fixtures

Light fixtures may have to be removed or detached and suspended (bailing wire works well) to gain access to ACM. Before beginning this task, the electrical supply should be shut off. Light fixtures should be wet-wiped before they are removed from the area. If it is not feasible to remove the light fixtures, they should be wet-wiped, then draped with plastic or completely enclosed.

STEP 11 — Securing the Work Area

When the work area is occupied, padlocks must be removed to permit emergency escape routes. Arrows should be taped on the polyethylene-covered walls to indicate the location of exits. All entrances should be secured when removal operations are not in progress. Provisions must also be made to secure the decontamination station entrance when no one is on the job site. Security guards may be a reasonable precaution, depending on the nature of the project.

Nonessential personnel should not be permitted to enter the work area. An on-site job log should be maintained for recording who enters the work area and the time each person enters and exits the work zone.

Establishing a Decontamination Unit

The decontamination station is designed to allow passage to and from the work area during removal operations with minimal leakage of asbestos-containing dust to the outside. A typical unit consists of a clean room, a shower room, and an equipment room separated by airlocks. The airlocks are formed by overlapping two sheets of polyethylene at the exit of one room, and two sheets at the entrance to the next room with three feet of space between the barriers. There are various methods for constructing airlocks including a hatch type construction and a slit and cover design.

Materials used to construct a typical unit include 2-in. by 4-in. lumber for the frame, 1/4-in. to 1/2-in. plywood or 6-mil polyethylene for the walls, duct tape, staples, and nails. The floor should be covered with three layers of 6-mil polyethylene. The decontamination unit can be built in sections to allow for disassembly and reuse at another area of the building. The design of the decontamination station will vary with each project depending on the size of the crew and the physical constraints imposed by the facility.

Customized trailers which can be readily moved from one location to the next are also used as decontamination stations. These units typically cost $20,000 to $50,000, depending on the size and features. A company conducting work at many different locations would probably recover this initial investment over time.

Whether a decontamination station is constructed on-site or is in the form of a trailer, the basic design is the same. The major components and their uses are discussed below.

Clean Room. No asbestos-contaminated items should enter this room. Workers use this area to suit up, store street clothes, and don respiratory protection on their way to the work area, and to dress in clean clothes after showering. This room should ideally be furnished with benches, lockers for clothes and valuables, and nails for hanging respirators.

Shower Room. Workers pass through the shower room on their way to the removal area, and use the showers on their way out after leaving contaminated clothing in the equipment room. Although most job specifications require only a single shower head, installation of multiple showers may be time- and cost-effective if the work crew is large. Shower wastewater should be collected and treated as ACM or filtered before disposal into the sanitary sewer. State and local requirements on methods of shower wastewater disposal vary. For example, Alabama, Georgia, Maryland, and New Jersey each have written specifications for handling shower wastewater.

Equipment Room. This is a contaminated area where equipment, boots or shoes, hardhats, goggles, and any additional contaminated work clothes are stored. Workers place disposable clothing such as coveralls, booties, and hoods in bins before leaving this area for the shower room. Respirators are worn until workers enter the shower and thoroughly soak them with water. The equipment room may require cleanup several times daily to prevent asbestos material from being tracked into the shower and clean rooms.

Waste Load-Out Area. This is an area separate from the decontamination unit which is used as a short-term storage area for bagged waste and as a port for transferring waste to the truck. An enclosure can be constructed to form an airlock between the exit of the load-out area and an enclosed truck.

The outside of the waste containers should be free of all contaminated material before removal from the work area. Gross contamination should be wiped or scraped off containers before they are placed in the load-out area. Any remaining contamination should be removed by wet-wiping or the bagged material can be placed in a second clean bag. To save cleanup time, fiber drums can be covered with an outside bag of polyethylene before they are taken into the work area; the bag can be removed before taking the drum into the load-out area.

Materials and Equipment List for Preparation of the Work Area and Establishing the Decontamination Station

Polyethylene Sheeting Material

Used to seal off work areas and items within work areas, protect surfaces in the work area other than those being altered, and construct decontamination and enclosure systems.

Types: 20' x 100' rolls, 4-mil thickness; 12' x 100' rolls, 60 lbs; 6-mil thickness, 20 lbs.

Duct Tape

Used to seam polyethylene sheets together; form airtight seal between polyethylene and walls; provide some support for vertical sheets.

Adhesive Spray

Used to seal seams; provide extra support to vertical sheets.

Furring Strips (cut into blocks)

Used to support vertical sheets of polyethylene.

Nails

Used to attach furring strips to top edge of polyethylene and then to the wall; construct the frame of the decontamination unit.

Staples and Staple Gun

Used to attach polyethylene to wood frame.

Retractable Razor Knives

Used to slice polyethylene and tape.

Warning Signs

Used to post entrances to building and decontamination unit.

Vacuum Cleaner Equipped with a HEPA Filter

Used to clean nonstationary items before removing them from the work area.

Ladders and/or Scaffolding

Carpentry Tools Such as Hammers, Saws, etc.

Prefab Shower Stalls or Materials for Shower Construction

REMOVAL OPERATIONS

Almost all asbestos removal projects involve wetting the asbestos with amended water. This may be spray-applied to exposed material or injected through holes punched into jacketing material with special pressure applications. The asbestos is then removed with hard scrapers, and all surfaces are brushed clean with nylon or fiber bristle brushes. Finally, a wet wipe with an encapsulant or lockdown solution completes the gross removal phase of the project. The decision to construct one large containment or create several small work areas is dictated by the nature of the project. When possible, it is often more cost-effective to create one large work area; however, sometimes it is desirable to divide up the work into smaller sections. For example, removing ACM from structural beams in a parking garage may be better accomplished with several narrow containments around the beams, thereby avoiding closing down the entire area as well as eliminating the need for a large cleanup area.

Creativity is also possible in this phase. In the above-mentioned parking area, for example, a contractor could create a 40-foot-long moveable containment area which can be slid along the floor. This could allow a removal crew to complete two 40-foot sections a day by moving the entire structure, result in substantial savings in material, and produce an easy-to-manage project. Further guidance on the removal phase can be found in Chapter 6 as well as in Appendix D.

Remove Gross Contamination from Floor Coverings or Remove Top Layer of Polyethylene If Two Layers Are Present

At this point, if two layers of 6-mil polyethylene have been used to cover the floor area, the sheets forming the top layer should be lightly misted and carefully folded inward to form compact bundles for bagging and disposal. Any visible contamination which leaked through to the inner floor layer should be removed (i.e., squeegeed, HEPA-vacuumed, wet-wiped). Excessive time should not be spent in cleaning the floor sheets, but any obvious contamination should be removed.

Conduct Visual Inspection of All Surface Areas; Reclean If Necessary

After these tasks have been accomplished, a thorough visual inspection of the area should be conducted. The inspector (building owner's representative) and the contractor's representative, usually the project supervisor, should check for visual contamination on the substrate from which the ACM has been removed, on ledges, on tops of doors, indented corners, and other areas which might catch falling material or contain residual material. A high-intensity flashlight will prove helpful during this inspection. As the inspector and job supervisor walk through the area, the inspection and recleaning process might be facilitated by recording on paper the items or areas which need additional cleaning. The contractor's representative is responsible for correcting any of the deficiencies noted during the inspection before beginning the next phase of work.

Perform Final Wipedown of Equipment and Remove from Work Area

After the work crew has completed recleaning the areas noted on the inspection list, the equipment should be thoroughly cleaned (gross contamination was removed earlier). Equipment should be wet-wiped, washed off in the shower at the waste load-out area, wrapped in polyethylene, or placed in plastic bags. Tools such as scrapers, utility knives, and brushes can be placed in buckets or pans (bottoms cut off fiberboard drums work well) and then sealed in plastic bags for transport to the next project. Brooms should be discarded or sealed in plastic bags. Equipment which is not needed for completion of the project should be removed from the work area. The negative air filtration unit remains in place and operational for the remainder of the cleanup operation until clearance samples are collected.

Apply Sealant to Substrate

The next phase of the project may include applying a sealant to the substrate and remaining plastic to "lock down" any tiny invisible fibrils which might remain. Also, the mist which occurs during application of the sealant aids in settling out and sticking down fibers which are still airborne.

Wait Overnight; Remove Polyethylene from Walls

An overnight waiting period (12-24 hr) should be provided after the sealant and/or sprayback material has been applied (or following recleaning after the inspection if no sealant is applied). This period allows the airborne materials to settle. The next day the polyethylene draped over lighting fixtures and covering the interior walls of the work area can be misted and carefully taken down, folded inward to form a bundle, and packaged for disposal. All coverings on doors, windows, and vents remain in place.

HEPA-Vacuum

After the walls are uncovered, the hard-to-reach places such as crevices around windows, doors, shelves, etc., can be cleaned using a vacuum equipped with a HEPA filter. On some projects, contractors may elect to vacuum all surface areas, beginning at the top of the wall and working downward. The HEPA filter retains the tiny fibers which could pass through a standard vacuum cleaner. HEPA vacuums are available with various canister sizes and horsepower motors. Some models have an available kit for converting a dry vacuum to a wet pick-up vacuum. Also models are available which use compressed air rather than the standard direct current. Twenty to thirty feet extension hoses are available for the larger vacuums.

Remove Polyethylene Floor Covering; Remove or Clean Carpet

After vacuuming of these areas is completed, the polyethylene floor covering is misted, each side is detached from the wall, and folded inward to form a compact bundle for bagging and disposal. If a carpet is in the work area and specified for removal (removal instead of cleaning is the preferred practice), workers should lightly mist the entire carpet before detaching it from the floor and rolling it up. Once the carpet is rolled up, it can be wrapped with 6-mil poly, sealed with duct tape, and labeled for disposal. A note of caution: in some locations, carpet may be stuck to the floor with a glue which does not readily separate from the flooring. As the carpet is taken up, some portions of the backing may tear away and remain stuck to the floor. Several unplanned additional man-hours may be required to pry or scrape up the glue-carpet spots which are left after the carpet is removed. Also, tearing of the carpet material may elevate fiber counts in air samples analyzed by phase contrast microscopy.

HEPA-Vacuum

After the floor area is uncovered, corners and crevices can be cleaned with a HEPA vacuum.

Wet-Wipe Walls

Next, the walls are wet-wiped and the floors are mopped (or if the carpet is left in place, it should be thoroughly vacuumed with a HEPA-filtered unit). Workers begin in the areas farthest away from the negative air filtration units and use amended water to wet-wipe all exposed surfaces (excluding the substrate from which the asbestos material was removed). For best results, workers should use cotton rags or lint-free paper towels which are disposed of after one use. Rinsing and reuse of towels may result in smearing asbestos fibers on the surfaces. Also, to avoid smearing of residual fibers, workers should wipe in one direction only. Paper towels should not be used to wipe down rough surfaces and should be discarded before they begin to deteriorate when used on smooth surfaces. Small "fibrous-looking" residue which may be deposited on surfaces as a result of fusing deteriorated paper towels could cause a problem during the final visual inspection.

Wet-Mop Floors

After the walls are wet-wiped, the floor is mopped with a clean mop head wetted with amended water. The water should be changed frequently. Waste water from the wet-wiping and wet-mopping operations is treated as asbestos-containing water and dumped in the shower drain or placed in a barrel for disposal.

Wait Overnight; Repeat Wet-Wipe and Wet-Mop Procedures

After the walls and other surfaces (shelves, ledges, etc.) have been wet-wiped and the floors have been mopped, activity in the area is stopped until the following day. The next day, the same wet-wiping and mopping procedures are repeated. If the carpet is left on the floor, it is HEPA-vacuumed again and steam cleaned. As an alternative to using amended water for the second wipedown, the cleaning efficiency may be increased by using a commercial cleaning product. Windows can also be cleaned with a commercial window cleaner.

Visual Inspection; Reclean If Necessary

The work area should be dry before the final visual inspection is conducted. The inspection is again conducted by the owner's representative and the job supervisor. All surfaces are carefully checked for visible contamination and any areas which need further cleaning are listed on paper. Be sure that ledges, tops of beams, and all hidden locations are also inspected for asbestos-containing dust.

Reinspect; Shut Off Negative Air Filtration Unit

After any designated areas have been recleaned, the inspector and job supervisor make a final walk-through to assure the items listed have been addressed. The negative air filtration units are shut off and the area is now ready for final clearance air monitoring.

Final Clearance Monitoring

Clearance monitoring is addressed in detail in the section on "Air Sampling Requirements." When the air sampling results indicate the airborne fiber concentration meets the criteria for clearance, the polyethylene can be removed from the vents, stationary objects such as water fountains, electrical outlets, etc., and any barriers can be removed. If the first set of air samples indicate airborne fiber concentrations in the area are above the specified clearance level, the area must be recleaned, followed again by clearance sampling. This cycle is repeated until results of airborne fiber concentrations indicate the clearance criteria have been attained.

After the area has been cleared for reoccupancy by unprotected personnel, remaining renovation can be initiated (i.e., painting walls, installing suspended ceiling, or laying carpet).

Cleaning Up the Decontamination Unit

The decontamination unit is lined with three layers of polyethylene on the floor and one or two layers on the walls (at a minimum, the walls of the equipment room should be lined). The top layer of floor poly in the equipment room should be removed at the time the top layer of floor poly in the work area is cleaned or removed, using the same procedures. This will minimize tracking contamination back into the work area. After cleanup is completed inside the work area, the polyethylene on the walls of the decontamination unit is misted and folded inward. Next, the remaining layers on the floor are removed in the same manner and packaged with the other poly for disposal. The walls should be visually checked for contamination and wet-wiped if necessary. The decontamination unit can now be disassembled for transport.

Cleaning Up the Enclosed Truck

During the last disposal run to the landfill, after the truck has been emptied of all waste materials, the polyethylene lining the inside of the truck is misted with amended water and carefully removed. Good practice should include wet-wiping the floor of the truck at this time. The polyethylene removed from the truck interior and the protective clothing worn by workmen conducting disposal are bagged for disposal and placed with the other materials at the dump site.[16]

8
Health and Safety

HEALTH HAZARD DATA

Asbestos, tremolite, anthophyllite, and actinolite can cause disabling respiratory disease and various types of cancers if the fibers are inhaled. Inhaling or ingesting fibers from contaminated clothing or skin can also result in these diseases. The symptoms of these diseases generally do not appear for 20 or more years after initial exposure.

Exposure to asbestos, tremolite, anthophyllite, and actinolite has been shown to cause lung cancer, mesothelioma, and cancer of the stomach and colon. Mesothelioma is a rare cancer of the thin membrane lining of the chest and abdomen. Symptoms of mesothelioma include shortness of breath, pain in the walls of the chest, and/or abdominal pain.

Respirators and Protective Clothing

Respirators

Workers are required to wear a respirator when performing tasks that result in asbestos, tremolite, anthophyllite, and actinolite exposure that exceeds the permissible exposure limit (PEL) of 0.2 f/cm^3. These conditions can occur while the employer is in the process of installing engineering controls to reduce asbestos, tremolite, anthophyllite, and actinolite exposure or where engineering controls are not feasible to reduce asbestos, tremolite, anthophyllite, and actinolite exposure. Air-purifying respirators equipped with high-efficiency particulate air (HEPA) filters can be used where airborne asbestos, tremolite, anthophyllite, and actinolite fiber concentrations do not exceed 2 f/cm^3; otherwise, air-supplied, positive-pressure, full-facepiece respirators must be used. Disposable respirators or dust masks are not permitted to be used for asbestos, tremolite, anthophyllite, and actinolite work. For effective protection, respirators must fit the face and head snugly. The employer is required to conduct fit tests when the worker is first assigned a respirator and every six months thereafter. Respirators should not be loosened or removed in work situations where their use is required.

Protective Clothing

Workers are required to wear protective clothing to prevent contamination of the skin in work areas where asbestos, tremolite, anthophyllite, and actinolite fiber concentrations exceed the PEL of 0.2 f/cm^3. Where protective clothing is required, the employer must provide the worker with clean garments. Unless employees are working on a large asbestos, tremolite, anthophyllite, and actinolite removal or demolition project, the employer must also provide a change room and separate lockers for workers' street clothes and contaminated work clothes. If employees are working on a large asbestos, tremolite, anthophyllite, and actinolite removal or demolition project, and where it is feasible to do so, the employer must provide a clean room, shower, and decontamination room contiguous to the work area. When leaving the work area, workers must remove contaminated clothing before proceeding to the shower. If the shower is not adjacent to the work area, workers must vacuum their clothing before proceeding to the change room and shower. To prevent inhaling fibers in contaminated change rooms and showers, workers must leave their respirator on until leaving the shower and entering the clean change room.

Disposal Procedures and Cleanup

Wastes that are generated by processes where asbestos, tremolite, anthophyllite, and actinolite are present include:

- empty asbestos, tremolite, anthophyllite, and actinolite shipping containers
- process wastes such as cuttings, trimmings, or reject material
- housekeeping waste from sweeping or vacuuming
- asbestos fireproofing or insulating material that is removed from buildings
- asbestos-containing building products removed during building renovation or demolition
- contaminated disposable protective clothing

Empty shipping bags can be flattened under exhaust hoods and packed into airtight containers for disposal. Empty shipping drums are difficult to clean and should be sealed.

Vacuum bags or disposable paper filters should not be cleaned but should be sprayed with a fine water mist and placed in a labeled waste container.

Process waste and housekeeping waste should be wetted with water, or a mixture of water and surfactant, prior to packaging in disposable containers.

Asbestos-containing material removed from buildings must be disposed of in leak-tight 6-mil-thick plastic bags, plastic-lined cardboard containers, or plastic-lined metal containers. These wastes, which are removed while wet, should be sealed in containers before they dry out to minimize the release of asbestos, tremolite, anthophyllite, and actinolite fibers during handling.

Access to Information

Each year the employer is required to inform the worker of the information contained in this standard and appendices for asbestos. In addition, the employer must instruct the worker in the proper work practices for handling asbestos-containing materials (ACM) and in the correct use of protective equipment.

The employer is required to determine whether a worker is being exposed to asbestos. The worker has the right to observe employee measurements and to record the results obtained. The employer is required to inform the worker of his/her exposure, and if the worker is exposed above the permissible limit, the employer is required to inform the worker of the actions that are being taken to reduce his/her exposure to within the permissible limit.

The employer is required to keep records of employee exposures and medical examinations. These exposure records must be kept for at least 30 years. Medical records must be kept for the period of the worker's employment plus 30 years.

The employer is required to release a worker's exposure and medical records to his/her physician, or designated representative, upon the worker's written request.

MEDICAL SURVEILLANCE GUIDELINES FOR ASBESTOS, TREMOLITE, ANTHOPHYLLITE, AND ACTINOLITE (as published by OSHA)[3]

Clinical evidence of the adverse effects associated with exposure to asbestos, tremolite, anthophyllite, and actinolite is present in the form of several well-conducted epidemiological studies of occupationally exposed workers, family contacts of workers, and persons living near asbestos, tremolite, anthophyllite, and actinolite mines. These studies have shown a definite association between exposure to asbestos, tremolite, anthophyllite, and actinolite and an increased incidence of lung cancer, pleural and peritoneal mesothelioma, gastrointestinal cancer, and asbestosis. The latter is a disabling fibrotic lung disease that is caused only by exposure to asbestos. Exposure to asbestos, tremolite, anthophyllite, and actinolite has also been associated with an increased incidence of esophageal, kidney, laryngeal, pharyngeal, and buccal cavity cancers. As with other known chronic occupational diseases, disease associated with asbestos, tremolite, anthophyllite, and actinolite generally appears about 20 years following the first occurrence of exposure. There are no known acute effects associated with exposure to asbestos, tremolite, anthophyllite, and actinolite.

Epidemiological studies indicate that the risk of lung cancer among exposed workers who smoke cigarettes is greatly increased over the risk of lung cancer among nonexposed smokers or exposed nonsmokers. These studies suggest that cessation of smoking will reduce the risk of lung cancer for a person exposed to asbestos, tremolite, anthophyllite, and actinolite, but will not reduce it to the same level of risk as that existing for an exposed worker who has never smoked.

Signs and Symptoms of Exposure-Related Disease

The signs and symptoms of lung cancer or gastrointestinal cancer induced by exposure to asbestos, tremolite, anthophyllite, and actinolite are not unique, except that a chest X-ray of an exposed patient with lung cancer may show pleural plaques, pleural calcification, or pleural fibrosis. Symptoms characteristic of mesothelioma include shortness of breath, pain in the walls of the chest, or abdominal pain. Mesothelioma has a much longer latency period compared with lung cancer (40 years, versus 15 to 20 years) and mesothelioma is therefore more likely to be found among workers who were first exposed to asbestos at an early age. Mesothelioma is always fatal.

Asbestosis is pulmonary fibrosis caused by the accumulation of asbestos fibers in the lungs. Symptoms include shortness of breath, coughing, fatigue, and vague feelings of sickness. When the fibrosis worsens, shortness of breath occurs even at rest. The diagnosis of asbestosis is based on a history of exposure to asbestos, the presence of characteristic radiologic changes, end-inspiratory crackles (rales), and other clinical features of fibrosing lung disease. Pleural plaques and thickening are observed on X-rays taken during the early stages of the disease. Asbestosis is often a progressive disease even in the absence of continued exposure, although this appears to be a highly individualized characteristic. In severe cases death may be caused by respiratory or cardiac failure.

Surveillance and Preventive Considerations

As noted above, exposure to asbestos, tremolite, anthophyllite, and actinolite has been linked to an increased risk of lung cancer, mesothelioma, gastrointestinal cancer, and asbestosis among occupationally exposed workers. Adequate screening tests to determine an employee's potential for developing serious chronic diseases such as cancer from exposure to asbestos, tremolite, anthophyllite, and actinolite do not presently exist. However some tests, particularly chest X-rays and pulmonary function tests, may indicate that an employee has been overexposed to asbestos, tremolite, anthophyllite, and actinolite, increasing his or her risk of developing exposure-related chronic diseases. It is important for the physician to become familiar with the operating conditions in which occupational exposure to asbestos, tremolite, anthophyllite, and actinolite is likely to occur. This is particularly important in evaluating medical and work histories and in conducting physical examinations. When an active employee has been identified as having been overexposed to asbestos, tremolite, anthophyllite, and actinolite, measures taken by the employer to eliminate or mitigate further exposure should also lower the risk of serious long-term consequences.

The employer is required to institute a medical surveillance program for all employees who are or will be exposed to asbestos, tremolite, anthophyllite, and actinolite at or above the action level (0.1 f/cm^3 of air) for 30 or more days per year and for all employees who are assigned to wear a negative-pressure respirator. All examinations and procedures must be performed by or under the supervision of a licensed physician at a reasonable time and place, and at no cost to the employee.

Although broad latitude is given to the physician in prescribing specific tests to be included in the medical surveillance program, OSHA requires inclusion of the following elements in the routine examination:

1. medical and work histories, with special emphasis directed to symptoms of the respiratory system, cardiovascular system, and digestive tract
2. completion of the respiratory disease questionnaire contained in Appendix D of OSHA Part 1910

3. a physical examination, including a chest roentgenogram and pulmonary function test that includes measurement of the employee's forced vital capacity (FVC) and forced expiratory volume at one second (FEV 1)
4. any laboratory or other test that the examining physician considers, by sound medical practice, to be necessary. The employer is required to make the prescribed tests available at least annually to those employees covered, and more often than specified if recommended by the examining physician, and upon termination of employment.

The employer is required to provide the physician with the following information: a copy of the OSHA standard and appendices; a description of the employee's duties as they relate to asbestos exposure; the employee's representative level of exposure to asbestos, tremolite, anthophyllite, and actinolite; a description of any personal protective and respiratory equipment used; and information from previous medical examinations of the affected employee that is not otherwise available to the physician. Making this information available to the physician will aid in the evaluation of the employee's health in relation to assigned duties and fitness to wear personal protective equipment, if required.

The employer is required to obtain a written opinion from the examining physician containing the results of the medical examination; the physician's opinion as to whether the employee has any detected medical conditions that would place the employee at an increased risk of exposure-related disease; any recommended limitations on the employee or on the use of personal protective equipment; and a statement that the employee has been informed by the physician of the results of the medical examination and of any medical conditions related to asbestos, tremolite, anthophyllite, and actinolite exposure that require further explanation or treatment. This written opinion must not reveal specific findings or diagnoses, and a copy of the opinion must be provided to the affected employee.

MEDICAL SURVEILLANCE

Contractors are required as employers to provide, at no charge to the employee (if exposed to asbestos), a physical examination by a qualified physician. There are specific items that these physicals must entail (refer to OSHA 29 CFR 1910.1001). The contractor may use the results of these physicals to screen potential employees who may have had previous exposures to asbestos. If possible, the contractor should avoid hiring a heavy smoker as a removal worker, or anyone else who would naturally be at an increased risk from previous exposure.

EMPLOYEE TRAINING

Any workers who will be in or around an asbestos abatement work area should, as a minimum, be advised of the hazards associated with asbestos exposure, be trained in how to adequately protect themselves from exposure during the course of the project, and be trained in correct job procedures for each of their positions. Training for the use of respirators is required by OSHA. A good way of documenting that this training has taken place is to develop a formal training session at which attendance is a mandatory condition of employment. After the training is complete, a written test should be administered. Those who pass the exam should be permitted to proceed with work, and those who fail should be held back, counseled as to why they failed, and subsequently retrained. A typical training program will be at least six to eight hours long, and for more complex jobs should last two days.

A good, in-depth training program should cover many concepts dealing with the various aspects of asbestos abatement projects, including background information on asbestos. Employees should be told what it is and where it comes from. Also, they should be informed as to how asbestos was used and why. An architect, or someone else with a good technical background, may be best suited to present this part of the training program.

The next phase of the training should be an outline of the dangers or health hazards associated with exposure to asbestos fibers. Someone with a good understanding of the medical hazards associated with breathing asbestos should give this part of the training session. It is important that not only the health hazards be discussed, but also how the fibers enter the body, and what happens once they are contained inside the lungs. Fiber size, visibility, and settling times are all important information. A film or slides may be helpful in illustrating these points.

After employees are made aware of the health hazards associated with asbestos exposure, the next phase of training should be on what they can do to protect themselves from this exposure (i.e., work practices and personal protective equipment). This training should include step-by-step instruction on how to perform each task associated with their jobs (i.e., glovebagging, wetting and scraping, etc.). Also, training should include a comprehensive review of the use of respiratory protection, including the following aspects:

- how to put on and take off the respirator
- cleaning and maintenance of respirators
- inspection of respirators
- fit testing of respirators
- uses and limitations of different types of respirators
- hands-on experience (look at various parts)

Note: The training requirements of an effective respiratory protection program are addressed in the section entitled "Respiratory Protection."

It is also important that workers be properly trained in the use of protective clothing. They should be made aware of its limitations, and how it should be used to optimize the protection factor.

The next phase of the training program should be a discussion of all applicable EPA and OSHA regulations regarding asbestos abatement projects. Also, there may be certain state or local regulations of which employees need to be aware. This part of the program should not be extremely detailed; rather, it should provide the employees with a good understanding of what they should or should not do when conducting removal of ACM. It should be emphasized that the main concern is the safety and health of the workers, rather than simply the concern of receiving a citation for a violation.

The fifth phase of employee training should deal with proper techniques for sealing off the work area. In this section, employees will be instructed on what to look for before sealing off the work area, and also on how to construct a safe and effective enclosure. Employees should first be made aware of what an HVAC system is, and how it affects the air movement through an area. They should also be instructed on how to shut the system down and seal off outlets and inlets so that airborne fibers will not be drawn into it. Employees should then be instructed in proper techniques for erecting plastic barriers, draping the walls, floors, and furniture with 6-mil polyethylene. This also includes construction of airlocks and change rooms, in addition to posting appropriate warning signs, etc. Also, it is important to inform employees that if a puncture develops in the polyethylene enclosure while the work area is active, they should stop work and immediately seal the leak.

Following the session on sealing off the work area, workers should be trained in how to effectively confine and minimize airborne fiber generation. This can best be accomplished through proper use of wet methods (i.e., spray the ACM with amended water). Workers should also be informed at this time that different forms of asbestos will react differently to the application of water. For example, chrysotile will typically accept water, while amosite is generally more resistant to wetting. Therefore, employees will have to take appropriate protective measures, since airborne fiber concentrations will be potentially higher when a removal job involves amosite. Employees should be instructed in methods of misting the air with water, and also in the proper methods of using the HEPA vacuum. Additionally, the function of negative air units should be outlined and employees made aware of the need to ensure that these units are kept running so that if a rupture occurred in the enclosure, fiber leakage would be minimized.

A very important aspect of employee training that is often taken lightly is the recognition and control of safety and health hazards (other than asbestos) in an asbestos abatement work area. Proper training can help reduce employee injuries and lost-time accidents. Subject areas that should be covered in this part of the session include the proper use of scaffolding, how to recognize and/or eliminate trip/slip hazards, the proper use of ladders, the identification of any electrical hazards, and how to avoid heat stress/heat stroke situations.

The next phase of employee training should entail cleaning up the work area. This cleaning will take place after gross removal has occurred and all residual debris is ready to be disposed of. Wet cleaning techniques should be reviewed (wetting the waste and collecting it off the floor). Settling times should also be discussed.

Correct disposal of asbestos-containing debris is also an important aspect of an abatement employee training program (specifically for employees who will be directly involved with disposal operations). This part of the program should include discussions on the need to place the wetted waste in appropriately labeled 6-mil polypropylene bags. These bags should then be placed in airtight fiberboard drums before being loaded into the enclosed truck to be taken to the landfill. It is important that any employees who might be involved in some way with this type of operation be made aware of the proper procedures for carrying out these waste disposal activities, and of the protective equipment required.

Another important aspect that an employee training program should include is information on final inspection/air sampling, and an explanation of why it is important. The reason employees need to be aware of what the final inspections will entail is because when they finish work in a certain area, they can conduct a fairly thorough visual inspection themselves.

Employees should also be informed as to why air sampling is being conducted, and what the results mean. Employees should be told that they may be asked to wear a personal air-sampling pump while they are performing their job, so that the fiber levels they are exposed to can be closely monitored. They should be requested to cooperate with the industrial hygienist when it is their turn to wear the sampling equipment. It must also be emphasized to the employees that they must not tamper with the sampling equipment they are wearing, since the results will indicate the level of airborne fibers to which they are being exposed.

Another aspect that the training program should cover is an explanation of the personnel decontamination sequence. This should cover procedures to be followed when beginning or finishing a shift of asbestos abatement work. When beginning a shift, employees should be instructed to enter the clean room first, put on their protective equipment/clothing, and proceed through the shower area and into the dirty equipment room before entering the work area. When finishing a shift, employees should be instructed to enter the dirty equipment room first, remove all of their protective clothing (except respirators), and then take showers. Respirators should be removed and washed while the employee is in the shower. After the employees complete the shower, they may then go to the clean room to change back into their street clothes. Employees should also be instructed that in emergency situations, the emergency will probably override the potential of adjacent area contamination, and good judgment should be used if someone needs to get out of the work enclosure very quickly (employee has a heart attack, fire, etc.). Emergency breakthrough points in the polyethylene enclosure should be clearly marked so that they will be easily accessible.

Once all of this formal classroom training is completed, ample time should be provided for employees to participate in hands-on training or workshops. Demonstration in these workshops should include proper techniques for glovebagging, wet and scrape methods, constructing work area enclosures, personal protective equipment, etc. This will be the most effective way to illustrate exactly how these typical asbestos abatement procedures should be conducted.

Throughout the training program, slides, videotapes, and handouts should be utilized when possible. A mixture of training techniques results in better learning. Also, employees with prior experience in asbestos abatement may have valuable input during this session. These comments should be encouraged by the instructor, and any misinformation should be immediately corrected without "putting the person down." It is important for the instructors not to get overly technical in any one area. At the end of the program, the written test should be administered. Results of these tests should be used to spot areas where employees may need further training. Also, at this time, employees should sign a form indicating that they have received training. The tests and the forms should then be placed in a file so that there will be documentation that employees were trained appropriately.

HEAT STRESS AND OTHER PHYSIOLOGICAL FACTORS (FROM NIOSH PUB.#85-115[17])

Personal Protective Equipment

Wearing personal protective equipment (PPE) puts a worker at considerable risk of developing heat stress. This can result in health effects ranging from transient heat fatigue to serious illness or death. Heat stress is caused by a number of interacting factors, including environmental conditions, clothing, workload, and the individual characteristics of the worker. Because heat stress is probably one of the most common (and potentially serious) illnesses at hazardous waste sites, regular monitoring and other preventive precautions are vital.

Individuals vary in their susceptibility to heat stress. Factors that may predispose someone to heat stress include:

- lack of physical fitness
- lack of acclimatization
- age
- dehydration
- obesity
- alcohol and drug use
- infection
- sunburn
- diarrhea
- chronic disease

Reduced work tolerance and the increased risk of excessive heat stress is directly influenced by the amount and type of PPE worn. Personal protective equipment adds weight and bulk, severely reduces the body's access to normal heat-exchange mechanisms (evaporation, convection, and radiation) and increases energy expenditure. Therefore, when selecting PPE, each item's benefit should be carefully evaluated in relation to its potential for increasing the risk of heat stress. Once PPE is selected, the safe duration of work/rest periods should be determined based on the:

- anticipated work rate
- ambient temperature and other environmental factors
- type of protective ensemble
- individual worker characteristics and fitness

Monitoring

Because the incidence of heat stress depends on a variety of factors, all workers, even those not wearing protective equipment, should be monitored.

For workers wearing permeable clothing (e.g., standard cotton or synthetic work clothes), follow recommendations for monitoring requirements and suggested work/rest schedules in the current American Conference of Governmental Industrial Hygienists' (ACGIH) Threshold Limit Values for Heat Stress. If the actual clothing worn differs from the ACGIH standard ensemble in insulation value and/or wind and vapor permeability, change the monitoring requirements and work/rest schedules accordingly.

For workers wearing semipermeable or impermeable encapsulating ensembles, the ACGIH standard cannot be used. For these situations, workers should be monitored when the temperature in the work area is above 70°F (21°C).[6]

To monitor the worker, measure:

- Heart rate. Count the radial pulse during a 30-second period as early as possible in the rest period. If the heart rate exceeds 110 beats/min at the beginning of the rest period, shorten the next work cycle by one-third and keep the rest period the same. If the heart rate still exceeds 110 beats/min at the next rest period, shorten the following work cycle by one-third.
- Oral temperature. Use a clinical thermometer (3 min under the tongue) or similar device to measure the oral temperature at the end of the work period (before drinking). If oral temperature exceeds 99.6°F (37.6°C), shorten the next work cycle by one-third without changing the rest period. If oral temperature still exceeds 99.6°F (37.6°C) at the beginning of the next rest period, shorten the following work cycle by

one-third. Do not permit a worker to wear a semipermeable or impermeable garment when his/her oral temperature exceeds 100.6°F (38.1°C).

- Body water loss, if possible. Measure weight on a scale accurate to 0.25 lb at the beginning and end of each work day to see if enough fluids are being taken to prevent dehydration. Weights should be taken while the employee wears similar clothing or, ideally, is nude. The body water loss should not exceed 1.5% total body weight in a work day.

Initially, the frequency of physiological monitoring depends on the air temperature adjusted for solar radiation and the level of physical work. The length of the work cycle will be governed by the frequency of the required physiological monitoring.

Prevention

Proper training and preventive measures will help avert serious illness and loss of work productivity. Preventing heat stress is particularly important because once someone suffers from heat stroke or heat exhaustion, that person may be predisposed to additional heat injuries. To avoid heat stress, management should take the following steps:

1. Adjust work schedules:

   - Modify work/rest schedules according to monitoring requirements.
   - Mandate work slowdowns as needed.
   - Rotate personnel: alternate job functions to minimize overstress or overexertion at one task.
   - Add additional personnel to work teams.
   - Perform work during cooler hours of the day, if possible, or at night if adequate lighting can be provided.

2. Provide shelter (air-conditioned, if possible) or shaded areas to protect personnel during rest periods.
3. Maintain workers' body fluids at normal levels. This is necessary to ensure that the cardiovascular system functions adequately. Daily fluid intake must approximately equal the amount of water lost in sweat, i.e., 8 fl oz (0.23 L) of water must be ingested for approximately every 8 oz (0.23 kg) of weight lost. The normal thirst mechanism is not sufficiently sensitive to ensure that enough water will be drunk to replace lost sweat. When heavy sweating occurs, encourage the worker to drink more. The following strategies may be useful:

   - Maintain water temperature at 50 to 60°F (10 to 15.6°C).
   - Provide small disposable cups that hold about 4 oz (0.1 L).
   - Have workers drink 16 oz (0.5 L) of fluid (preferably water or dilute drinks) before beginning work.
   - Urge workers to drink a cup or two every 15 to 20 min, or at each monitoring break. A total of 1 to 1.6 gal (4 to 6 L) of fluid per day are recommended, but more may be necessary to maintain body weight.
   - Weigh workers before and after work to determine if fluid replacement is adequate.

4. Encourage workers to maintain an optimal level of physical fitness:

   - Where indicated, acclimatize workers to site work conditions (temperature, protective clothing, and workload).
   - Urge workers to maintain normal weight levels.

5. Provide cooling devices to aid natural body heat exchange during prolonged work or severe heat exposure. Cooling devices include:

   - field showers or hose-down areas to reduce body temperature and/or to cool off protective clothing
   - cooling jackets, vests, or suits

Table 2. Suggested Frequency of Physiological Monitoring for Fit and Acclimatized Workers

Adjusted Temperature	Normal Work Ensemble	Impermeable Ensemble
90 ° F (32.2 ° C) or above	45 min	15 min
87.5-90 ° F (30.8–32.2 ° C)	60 min	30 min
82.5-87.5 ° F (28.1–30.8 ° C)	90 min	60 min
77.5-82.5 ° F (25.3–28.1 ° C)	120 min	90 min
72.5-77.5 ° F (22.5–25.3 ° C)	150 min	120 min

6. Train workers to recognize and treat heat stress. As part of training, identify the signs and symptoms of heat stress.

Other Factors

Wearing PPE decreases worker performance as compared to an unequipped individual. The magnitude of this effect varies considerably, depending on both the individual and the PPE ensemble used. This section discusses the demonstrated physiological responses to PPE, the individual human characteristics that play a factor in these responses, and some of the precautionary and training measures that need to be taken to avoid PPE-induced injury.

The physiological factors that may affect worker ability to function using PPE include:

- physical condition
- level of acclimatization
- age
- gender
- weight

Table 2 shows suggested frequencies of physiological monitoring for fit and acclimatized workers at various temperatures.

Signs and Symptoms of Heat Stress

- Heat rash may result from continuous exposure to heat or humid air.
- Heat cramps are caused by heavy sweating with inadequate electrolyte replacement; signs and symptoms include muscle spasms and pain in the hands, feet, and abdomen.
- Heat exhaustion occurs from increased stress on various body organs, including inadequate blood circulation due to cardiovascular insufficiency or dehydration. Signs and symptoms include pale, cool, moist skin; heavy sweating; dizziness; nausea; and fainting.
- Heat stroke is the most serious form of heat stress. Temperature regulation fails and the body temperature rises to critical levels. Immediate action must be taken to cool the body before serious injury and death occur. Competent medical help must be obtained. Signs and symptoms are red, hot, usually dry skin; lack of or reduced perspiration; nausea; dizziness and confusion; strong, rapid pulse; and coma.

Physical Condition

Physical fitness is a major factor influencing a person's ability to perform work under heat stress. The more fit someone is, the more work they can safely perform. At a given level of work, a fit person, relative to an unfit person, will have:

- less physiological strain
- a lower heart rate
- a lower body temperature, which indicates less retained body heat (a rise in internal temperature precipitates heat injury)

- a more efficient sweating mechanism
- slightly lower oxygen consumption
- slightly lower carbon dioxide production

Level of Acclimatization

The degree to which a worker's body has physiologically adjusted or acclimatized to working under hot conditions affects his or her ability to work. Acclimatized individuals generally have lower heart rates and body temperatures than unacclimatized individuals, and sweat sooner and more profusely. This enables them to maintain lower skin and body temperatures at a given level of environmental heat and work loads than unacclimatized workers. Sweat composition also becomes more dilute with acclimatization, which reduces salt loss. Acclimatization can occur after just a few days of exposure to a hot environment. The National Institute of Occupational Safety and Health recommends a progressive six-day acclimatization period for the unacclimatized worker before allowing him/her to do full work on a hot job. Under this regimen, the first day of work on site is begun using only 50% of the anticipated work load and exposure time, and 10% is added each day through day six. With fit or trained individuals, the acclimatization period may be shortened two or three days. However, workers can lose acclimatization in a matter of days, and work regimens should be adjusted to account for this.

When enclosed in an impermeable suit, fit acclimatized individuals sweat more profusely than unfit or unacclimatized individuals and may therefore actually face a greater danger of heat exhaustion due to rapid dehydration. This can be prevented by consuming adequate quantities of water.

Age

Generally, maximum work capacity declines with increasing age, but this is not always the case. Active, well-conditioned seniors often have performance capabilities equal to or greater than young sedentary individuals. However, there is some evidence, indicated by lower sweat rates and higher body core temperatures, that older individuals are less effective in compensating for a given level of environmental heat and work loads. At moderate thermal loads, however, the physiological responses of "young" and "old" are similar and performance is not affected.

Age should not be the sole criterion for judging whether or not an individual should be subjected to moderate heat stress. Fitness level is a more important factor.

Gender

The literature indicates that females tolerate heat stress at least as well as their male counterparts. Generally, a female's work capacity averages 10 to 30% less than that of a male. The primary reasons for this are the greater oxygen-carrying capacity and the stronger heart in the male. However, a similar situation exists as with aging: not all males have greater work capacities than all females.

Weight

The ability of a body to dissipate heat depends on the ratio of its surface area to its mass (surface area/weight). Heat loss (dissipation) is a function of surface area and heat production is dependent on mass. Therefore, heat balance is described by the ratio of the two.

Since overweight individuals (those with a low ratio) produce more heat per unit of surface area than thin individuals (those with a high ratio), overweight individuals should be given special consideration in heat stress situations. However, when wearing impermeable clothing, the weight of an individual is not a critical factor in determining the ability to dissipate excess heat.

9
Regulations

SAMPLE STATE REGULATION
(Based on the Recent Ohio Standard)[18]

Certification of Asbestos Hazard Abatement Contractors

A. The director shall determine which asbestos hazard abatement contractors are qualified to carry out the asbestos hazard abatement activities referred to in paragraph (1) of division (b) of section 506 of the ASHAA, based upon the criteria contained in this rule, and shall certify contractors who meet the criteria. Any asbestos hazard abatement contractor may apply to the director for certification on forms provided by the director. The director shall review any such applications and, within 30 days of receipt, if an application contains insufficient information for the director to make this determination, the director may request additional information from the applicant, in which case the director shall grant or deny the application within fifteen days of receipt of the necessary information. Any certification granted in accordance with this rule shall expire on April 1, 1986.

B. To become certified under this rule, an asbestos hazard abatement contractor shall provide documentation that he or she meets the following criteria:

1. Satisfactory completion of a training course in the removal and abatement of asbestos hazards, as described in paragraph (A) of rule 3701-34-05 of the administrative code;

2. Reliability in the performance of general contracting activities, as demonstrated through the submission of a list of persons who can attest to the quality of work performed by the contractor;

3. The applicant will provide to building owners, employees, and other relevant persons information that the director deems adequate regarding protection measures to be taken during the course of asbestos hazard abatement activities;

4. The applicant will provide to the applicant's employees or other persons utilized by the applicant for asbestos hazard abatement activities:

(a) Training in asbestos hazard abatement, to be evidenced by each of the persons being utilized by the applicant for asbestos hazard abatement activities having satisfactorily completed the training course described in paragraph (C) of rule 3701-34-05 of the administrative code;

(b) Personal protective equipment appropriate to the asbestos hazard abatement activity but no less than the personal protective equipment required by the asbestos regulations of the United States Occupational Safety and Health Administration, 29 CFR Section 1910.1001;

5. Prior experience in performing asbestos hazard abatement activities, as demonstrated through the submission of a list of at least three successfully completed prior contracts, including the names, addresses, and telephone numbers of building owners for whom the contracts were performed. Applicants also shall submit air monitoring data, if any, taken during and after completion of any of these projects in accordance with 29 CFR Section 1910.1001. The director may waive the experience requirement established by this paragraph if an inexperienced contractor can demonstrate exceptional qualifications under the other criteria of this rule;

6. Maintenance of satisfactory, written standard operating procedures, employee respiratory protection programs, overall work practices, and bulk and air-sampling procedures for asbestos hazard abatement activities, as demonstrated by submission of copies

of these procedures and plans, which shall include specific references to the medical monitoring and respiratory training programs of the United States Occupational Safety and Health Administration. The procedures and plans shall specify that the contractor will make available for viewing at all job sites copies of United States Occupational Safety and Health Administration regulations (29 CFR Section 1910.1001) and United States Environmental Protection Agency regulations (40 CFR Part 61, Subpart M) pertaining to asbestos hazard abatement activities;

7. Demonstrated competence in the performance of asbestos hazard abatement activities. In assessing an applicant's compliance with this criterion, the director shall consider the following information, which shall be provided in the application:

(a) Descriptions of any asbestos hazard abatement activities conducted by the applicant that have been prematurely terminated, including the circumstances surrounding the termination;

(b) A list of any contractual penalties that the applicant has paid for breach of or noncompliance with contract specifications for asbestos hazard abatement activities, such as overruns of completion time or liquidated damages;

(c) Identification of any citations levied against the applicant by any federal, state, or local government agencies for violations related to asbestos hazard abatement, including the name or location of the project, the date(s), and how the allegations were resolved; and

(d) A description, in detail, of all legal proceedings, lawsuits, or claims that have been filed or levied against the applicant or any of the applicant's past or present employees for asbestos-related activities, and how the allegations were resolved.

C. Asbestos hazard abatement contractors as defined in paragraph (H)(2) of Rule 3701-34-01 of the administrative code (asbestos hazard abatement coordinators for school districts utilizing their own employees for asbestos hazard abatement activities shall meet the criteria of paragraphs (B)(1), (B)(3), (B)(4), and (B)(6) of this rule in order to become certified under this rule.)

D. Nothing in this rule shall be construed to preclude an asbestos hazard abatement contractor who intends to perform asbestos hazard abatement activities exclusively in nonschool buildings from applying for certification under this rule.

ASBESTOS HAZARD ABATEMENT TRAINING COURSES (From Ohio Rules)

A. Any institution of higher education may apply to the director for approval of a course in asbestos hazard abatement as being sufficient for the training of asbestos hazard abatement contractors who wish to become certified. To be approved for this purpose, a training course shall meet the following criteria:

1. Provision of at least one and one-quarter clock-hours of instruction on each of the following topics or a combined total of at least twenty-two and one-half clock-hours of instruction that adequately addresses each of the topics:

- the history, properties, and uses of asbestos
- federal and state regulation of asbestos
- asbestos hazards and health effects
- building inspection, surveys, and testing
- sampling and analysis of bulk samples for asbestos
- sampling and analysis of atmospheric samples for asbestos
- asbestos health risk evaluation
- routine maintenance and minor repairs in asbestos environments
- removal and disposal of asbestos from nonstructural surfaces
- removal and disposal of asbestos from structural surfaces
- asbestos encapsulation, enclosure, and containment
- nonasbestos hazards
- specifications for projects in asbestos environments
- management and supervision of asbestos control projects
- worker protection
- asbestos liability/cost recovery alternatives
- recordkeeping responsibilities

- innovative approaches to the control of asbestos during removal projects

2. Records are maintained of persons who have attended or completed the course and of their attendance and completion dates, which information shall be provided to the director upon request;

3. Administration of an approved examination requiring a grade of 70% for satisfactory completion of the course, which grades shall be provided to the director upon request;

4. Issuance of a certificate to each student who satisfactorily completes the course, which certificate shall constitute certification by the department that the student has completed the course.

B. The asbestos hazard abatement training courses provided by the University of Cincinnati, the Georgia Institute of Technology, Tufts University, and the University of Kansas are deemed approved under paragraph A of this rule for the training of asbestos hazard abatement contractors.

C. Any person may apply to the director for approval of a course in asbestos hazard abatement as being sufficient for the training of workers who will undertake asbestos hazard abatement activities under the supervision of a certified contractor. Each applicant shall include with the application copies of the curriculum and of the examination for the course. To be approved for this purpose, a training course shall meet the following criteria:

1. Provision to each student of a total of at least five hours of instruction on the recognition of asbestos, the physical characteristics and uses of asbestos, the health hazards of asbestos, including the relationships between asbestos exposure, smoking, and various diseases, and the requirements, procedures, and standards established by government agencies for asbestos hazard abatement;

2. Provision to each student of a total of at least five hours of instruction on worker protection, including respiratory protection, protective clothing, safety equipment, air monitoring, medical surveillance, personal hygiene, work practices, including area preparation, decontamination, and waste disposal, and detailed description of respirators and their use and care, including protection factors, fitting and testing procedures, maintenance, and cleaning;

3. Provision to each student of at least 15 minutes of individual instruction concerning respirator fit tests and of the opportunity to use respirators;

4. Instruction is supervised by an industrial hygienist certified by the American Board of Industrial Hygiene or a person with other sufficient training, as determined by the director. Instruction shall be provided by persons with qualifications satisfactory to the director;

5. Records are maintained of persons who have attended or completed the course and of their attendance and completion dates, which information shall be provided to the director upon request;

6. Administration of an approved examination requiring a grade of 70% for satisfactory completion of the course, which grades shall be provided to the director upon request;

7. Issuance of a certificate to each student who satisfactorily completes the course, which certificate shall constitute certification by the director that the student has completed the course.

D. Priority in registration for training courses approved under paragraph A or C of this rule shall be afforded to state or local government employees with statutory or regulatory asbestos hazard abatement responsibility.

E. The denial of an application for approval or the revocation of an approval previously granted for a training course under this rule shall constitute an adjudication order under Chapter 119 of the Revised Code, and shall be effected subject to that chapter.

OSHA REGULATIONS SUMMARY: ASBESTOS, TREMOLITE, ANTHOPHYLLITE, AND ACTINOLITE[19]

Scope

Applies to all occupational exposures to asbestos, tremolite, anthophyllite, and actinolite in all industries.

Permissible Exposure Limit (PEL)

No employee is to be exposed to airborne concentrations of asbestos in excess of 0.2 f/cm^3 of air as an eight-hour time-weighted average (TWA).

Monitoring

Every employer shall perform initial monitoring on employees who are or may be exposed at or above the action level (0.1 f/cm^3). The employer may rely on objective data that demonstrates that asbestos is not capable of being released in concentrations in excess of the action level in lieu of sampling.

Monitoring is to be repeated at least every six months for employees whose exposures may reasonably be foreseen to exceed the action level. All samples taken to satisfy the monitoring requirements shall be personal samples.

The employer is required to notify the affected employees of the results of the monitoring within 15 days of the receipt of the results. This must be done in writing.

Regulated Areas

The employer is required to establish regulated areas wherever airborne concentrations of asbestos exceed the permissible exposure limit. These regulated areas must be demarcated in any manner that minimizes the number of persons who will be exposed to asbestos. Access to regulated areas is limited to authorized persons.

Each person entering a regulated area shall be provided with and required to use a respirator. Eating, drinking, smoking, the chewing of tobacco or gum, and the application of cosmetics are prohibited in the regulated area.

Methods of Compliance

The employer is required to institute engineering and work practice controls to reduce and maintain exposure to or below the PEL.

All hand-operated and power-operated tools which would produce or release fibers of asbestos must be provided with local exhaust ventilation. Where practical, asbestos is to be handled, mixed, applied, removed, or otherwise worked in a wet state.

Materials containing asbestos shall not be applied by spray methods. Compressed air is not to be used to remove asbestos unless it is used in conjunction with a ventilation system.

Compliance Program

Where the PEL is exceeded, the employer is required to establish and implement a written program to reduce employee exposures to or below the PEL. The employer shall not use employee rotation as a means of compliance with the PEL.

Respiratory Protection

The employer is required to provide and ensure the use of respirators where required. Appropriate respirators are to be provided at no cost to the employees.

Power air-purifying respirators (PAPRs) are to be provided if an employee chooses to use this type and it provides adequate protection. Where respirators are required, a written respirator program shall be instituted.

Employees who use filter respirators are to be permitted to change filters whenever an increase in breathing resistance is detected. Employees who wear respirators are to be allowed to leave the regulated area whenever necessary to wash their faces and respirator facepieces to prevent skin irritation.

Employees shall not be assigned to tasks which require the use of respirators if an examining physician determines that the employee will be unable to function normally wearing a respirator.

Quantitative or qualitative fit tests shall be performed on employees wearing negative-pressure respirators at the time of initial fitting and at least every six months thereafter.

Protective Work Clothing and Equipment

Where employees are exposed to asbestos above the PEL, the employer is required to provide to employees at no cost, appropriate protective work clothing and equipment.

The employer shall ensure that work clothing contaminated with asbestos be removed only in the change rooms provided. Employees are not permitted to remove contaminated work clothing from the change room.

Contaminated work clothing shall be stored in closed containers. Containers of contaminated protective equipment or work clothing are to be properly labeled.

Clean protective clothing is to be provided at least weekly to each affected employee.

Hygiene Facilities

The employer is required to provide clean change rooms for employees who are exposed to asbestos in excess of the PEL. Employees who are exposed to asbestos in excess of the PEL are required to shower at the end of the shift.

The employer is required to provide a positive-pressure, filtered-air lunchroom for employees whose asbestos exposure exceeds the PEL.

Communication of Hazards

Warning signs shall be provided and displayed at each regulated area. Warning labels complying with 1910.1200 shall be affixed to all asbestos-containing materials (ACM).

Employers who manufacture asbestos products are required to develop Material Safety Data Sheets that comply with 1910.1200.

The employer is required to institute a training program for employees who are exposed to airborne concentrations of asbestos at or above the action level.

Housekeeping

All surfaces shall be maintained as free as practicable of accumulations of asbestos. Surfaces are not to be cleaned by the use of compressed air.

Medical Surveillance

The employer must institute a medical surveillance program for employees who are or may be exposed to asbestos at or above the action level.

Before an employee is assigned to a job exposed to asbestos, a preplacement medical exam shall be provided. Periodic medical exams are to be provided annually. A termination of employment medical exam is to be given within 30 calendar days before or after the date of termination of employment to any employee who has been exposed to asbestos at or above the action level.

Recordkeeping

The employer is required to keep an accurate record of all measurements taken to monitor employee exposure to asbestos. The employer is required to establish and maintain an accurate record for each employee subject to medical surveillance. These records shall be maintained for at least 30 years.

Definitions

"Action level" means an airborne concentration of asbestos, tremolite, anthophyllite, actinolite, or a combination of these minerals, of 0.1 fiber per cubic centimeter (f/cm^3) of air calculated as an 8-hr time-weighted average.

"Asbestos" includes chrysotile, amosite, tremolite asbestos, anthophyllite asbestos, actinolite asbestos, and any of these minerals that have been chemically treated and/or altered.

"Fiber" means a particulate form of asbestos, tremolite, anthophyllite, or actinolite, 5 μm or longer, with a length-to-diameter ratio of at least 3 to 1.

"Employee exposure" means that exposure to airborne asbestos, tremolite, anthophyllite, actinolite, or a combination of these minerals that would occur if the employee were not using respiratory protective equipment.

"High-efficiency particulate air (HEPA) filter" means a filter capable of trapping and retaining at least 99.97% of 0.3 μm diameter mono-disperse particles.

"Regulated area" means an area established by the employer to demarcate areas where airborne concentrations of asbestos, tremolite, anthophyllite, actinolite, or a combination of these minerals exceeds, or can reasonably be expected to exceed, the permissible exposure limit.

"Tremolite, anthophyllite, or actinolite" means the nonasbestos form of these minerals, and any of these minerals that have been chemically treated and/or altered.

Permissible Exposure Limit (PEL)

The employer shall ensure that no employee is exposed to an airborne concentration of asbestos, tremolite, anthophyllite, actinolite, or a combination of these minerals in excess of 0.2 f/cm^3 of air as an 8-hr TWA.

Exposure Monitoring

Determinations of employee exposure will be made from breathing zone air samples that are representative of the 8-hr TWA of each employee.

Representative 8-hr TWA employee exposures shall be determined on the basis of one or more samples representing full-shift exposures for each shift for each employee in each job classification in each work area.

Initial Monitoring. Each employer who has a workplace or work operation covered by this standard shall perform initial monitoring of employees who are, or may reasonably be expected to be, exposed to airborne concentrations at or above the action level.

Where the employer has relied upon objective data that demonstrate that asbestos, tremolite, anthophyllite, actinolite, or a combination of these minerals is not capable of being released in airborne concentrations at or above the action level under the expected conditions of processing, use, or handling, no initial monitoring is required.

Monitoring Frequency and Patterns. After the initial determinations, samples shall be of such frequency and pattern as to represent with reasonable accuracy the levels of exposure of the employees. In no case shall sampling be at intervals greater than six months for employees whose exposures may reasonably be foreseen to exceed the action level.

Methods of Compliance

Engineering Controls and Work Practices. The employer shall institute engineering controls and work practices to reduce and maintain employee exposure to or below the exposure time prescribed in paragraph (c), except to the extent that such controls are not feasible. Wherever the feasible engineering controls and work practices that can be instituted are not sufficient to reduce employee exposure to or below the permissible exposure limit, the employer shall use them to reduce employee exposure to or below 0.5 f/cm^3 of air and shall supplement them by the use of any combination of respiratory protection, work practices, and feasible engineering controls that will reduce employee exposure to or below the permissible exposure limit for the following industries and operations: coupling cutoff in primary asbestos cement pipe manufacturing; sanding in primary and secondary asbestos cement sheet manufacturing; grinding in primary and secondary friction product manufacturing; carding and spinning in dry textile processes; and grinding and sanding in primary plastics manufacturing.

Compressed Air. Compressed air shall not be used to remove asbestos, tremolite, anthophyllite, or actinolite, or materials containing asbestos, tremolite, anthophyllite, or actinolite, unless the compressed air is used in conjunction with a ventilation system designed to capture the dust cloud created by the compressed air.

Compliance Program. Where the PEL is exceeded, the employer shall establish and implement a written program to reduce employee exposure to or below the limit by means of engineering and work practice controls.

Such programs shall be reviewed and updated as necessary to reflect significant changes in the status of the employer's compliance program.

Written programs shall be submitted upon request for examination and copying to the Assistant Secretary, the Director, affected employees, and designated employee representatives. The employer shall not use employee rotation as a means of compliance with the PEL.

Respirator Program. Where respiratory protection is required, the employer shall institute a respirator program in accordance with 29 CFR 1910.134. (See Table 3 for respiratory protection data).

The employer shall permit each employee who uses a filter respirator to change the filter elements whenever an increase in breathing resistance is detected and shall maintain an adequate supply of filter elements for this purpose.

Employees who wear respirators shall be permitted to leave the regulated area to wash their faces and respirator facepieces whenever necessary to prevent skin irritation associated with respirator use.

No employee shall be assigned to tasks requiring the use of respirators if, based upon his or her most recent examination, an examining physician determines that the employee will be unable to function normally wearing a respirator, or that the safety or health of the employee or other employees will be impaired by the use of a respirator. Such an employee shall be assigned to another job or given the opportunity to transfer to a different position whose duties he or she is able to perform with the same employer, in the same geographical area, and with the same seniority, status, and rate of pay the employee had just prior to such transfer, if such a different position is available.

SELECTED EDITED PORTIONS OF THE REGULATIONS: PART 1926.58; ASBESTOS, TREMOLITE, ANTHOPHYLLITE, AND ACTINOLITE[20]

Scope and Application

This section applies to all construction work as defined in 29 CFR 1910.12 (b) including but not limited to the following:

Table 3. Respiratory Protection for Asbestos Tremolite, Anthophyllite, and Actinolite Fibers

Airborne Concentrations of Asbestos, Tremolite, Anthophyllite, Actinolite, or a Combination of These Minerals		Required Respirator
Not in excess of 2 ft/cm^3 (10 × PEL)	1.	Half mask air-purifying respirator equipped with high-efficiency filters.
Not in excess of 10 ft/cm^3 (50 × PEL)	1.	Full facepiece air-purifying respirator equipped with high-efficiency filters.
Not in excess of 20 ft/cm^3 (100 × PEL)	1.	Any powered air-purifying respirator equipped with high-efficiency filters.
	2.	Any supplied-air respirator operated in continuous flow mode.
Not in excess of 200 ft/cm^3 (1000 × PEL)	1.	Full facepiece supplied air respirator operated in pressure demand mode.
Greater than 200 ft/cm^3 (> 1.000 × PEL) or unknown concentration	1.	Full facepiece supplied air respirator operated in pressure demand mode equipped with an auxiliary positive self-contained breathing apparatus.

Note: Respirators assigned for higher environmental concentrations may be used at lower concentrations. A high-efficiency filter means a filter that is at least 99.97% efficient against monodispersed particles of 0.3 μm or larger.

- demolition or salvage of structures where asbestos, tremolite, anthophyllite, or actinolite is present
- removal or encapsulation of materials containing asbestos, tremolite, anthophyllite, or actinolite
- construction, alteration, repair, maintenance, or renovation of structures, substrates, or portions thereof, that contain asbestos, tremolite, anthophyllite, or actinolite
- installation of products containing asbestos, tremolite, anthophyllite, or actinolite

"Demolition" means the wrecking or taking out of any load-supporting structural member and any related razing, removing, or stripping of asbestos, tremolite, anthophyllite, or actinolite products.

"Employee exposure" means that exposure to airborne asbestos, tremolite, anthophyllite, actinolite, or a combination of these minerals that would occur if the employee were not using respiratory protective equipment.

"Equipment room (change room)" means a contaminated room located within the decontamination area that is supplied with impermeable bags or containers for the disposal of contaminated protective clothing and equipment.

"High-efficiency particulate air (HEPA) filter" means a filter capable of trapping and retaining at least 99.97% of all monodispersed particles of 0.3 μm in diameter or larger.

"Regulated area" means an area established by the employer to demarcate areas where airborne concentrations of asbestos, tremolite, anthophyllite, actinolite, or a combination of these minerals exceed or can reasonably be expected to exceed the permissible exposure limit. The regulated area may take the form of a temporary enclosure or an area demarcated in any manner that minimizes the number of employees exposed to asbestos, tremolite, anthophyllite, or actinolite.

"Removal" means the taking out or stripping of asbestos, tremolite, anthophyllite, or actinolite, or materials containing asbestos, tremolite, anthophyllite, or actinolite.

"Renovation" means the modifying of any existing structure, or portion thereof, where exposure to airborne asbestos, tremolite, anthophyllite, or actinolite may result.

"Repair" means the overhauling, rebuilding, reconstructing, or reconditioning of structures or substrates where asbestos, tremolite, anthophyllite, or actinolite is present.

"Tremolite, anthophyllite, and actinolite" means the nonasbestos form of these minerals, and any of these minerals that have been chemically treated and/or altered.

Communication Among Employers

On multi-employer worksites an employer performing asbestos, tremolite, anthophyllite, or actinolite work requiring the establishment of a regulated area shall inform other employers on the site of the nature of the employer's work with asbestos, tremolite, anthophyllite, or actinolite and of the existence of and requirements pertaining to regulated areas.

Requirements for Asbestos Removal, Demolition, and Renovation Operations

Wherever it is feasible, the employer shall establish negative-pressure enclosures before commencing removal, demolition, and renovation operations.

The employer shall designate a competent person to perform or supervise the following duties: set up enclosure and ensure the integrity of the enclosure.

For small-scale, short-duration operations, such as pipe repair, valve replacement, installing electrical conduits, installing or removing drywall, roofing, and other general building maintenance or renovation, the employer is not required to comply with all the requirements of this section.

Exposure Monitoring

Each employer who has a workplace or work operation covered by this standard shall perform monitoring to determine accurately the airborne concentrations of asbestos.

Determinations of employee exposure shall be made from samples of breathing-zone air that are representative of the 8-hr TWA of each employee. Representative 8-hr TWA employee exposure shall be determined on the basis of one or more samples representing full-shift exposure for employees in each work area.

Periodic Monitoring Within Regulated Areas

The employer shall conduct daily monitoring that is representative of the exposure of each employee who is assigned to work within a regulated area.

Exception: When all of the employees in a regulated area are equipped with supplied-air respirators operated in the positive-pressure mode, the employer may dispense with the daily monitoring required by this paragraph.

Termination of Monitoring. If periodic monitoring required reveals that employee exposures, as indicated by statistically reliable measurements, are below the action level, the employer may discontinue monitoring for those employees whose exposures are represented by such monitoring.

Methods of Compliance

Engineering Controls and Work Practices. The employer shall use one or any combination of the following control methods to achieve compliance with the permissible exposure limit:

- local exhaust ventilation equipped with HEPA-filter dust collection systems
- general ventilation systems
- vacuum cleaners equipped with HEPA filters
- enclosure or isolation of processes producing asbestos, tremolite, anthophyllite, or actinolite dust
- use of wet methods, wetting agents, or removal encapsulants to control employee exposures during asbestos, tremolite, anthophyllite, or actinolite

handling, mixing, removal, cutting, application, and cleanup

- prompt disposal of wastes contaminated with asbestos, tremolite, anthophyllite, or actinolite in leak-tight containers or use of work practices or other engineering controls that the Assistant Secretary can show to be feasible

Wherever the feasible engineering and work practice controls described above are not sufficient to reduce employee exposure to or below the limit prescribed in text above, the employer shall use them to reduce employee exposure to the lowest levels attainable by these controls and shall supplement them by the use of respiratory protection that complies with the requirements of the above paragraphs of this section.

Prohibitions

High-speed abrasive-disc saws that are not equipped with appropriate engineering controls shall not be used for work related to asbestos, tremolite, anthophyllite, or actinolite.

Compressed air shall not be used to remove asbestos, tremolite, anthophyllite, or actinolite, or materials containing asbestos, tremolite, anthophyllite, or actinolite unless the compressed air is used in conjunction with an enclosed ventilation system designed to capture the dust cloud created by the compressed air.

Materials containing asbestos, tremolite, anthophyllite, or actinolite shall not be applied by spray methods.

Employee Rotation

The employer shall not use employee rotation as a means of compliance with the exposure limit prescribed in the above paragraphs.

Respiratory Protection

The employer shall provide respirators and ensure that they are used. Respirators shall be used in the following circumstances:

- during the interval necessary to install or implement feasible engineering and work practice controls
- in work operations such as maintenance and repair activities or other activities for which engineering and work practice controls are not feasible
- in work situations where feasible engineering and work practice controls are not yet sufficient to reduce exposure to or below the exposure limit
- in emergencies

Respirator Selection

Where respirators are used, the employer shall select and provide, at no cost to the employee, the appropriate respirator as specified in Table 3, and shall ensure that the employee uses the respirator provided.

The employer shall select respirators from among those jointly approved as being acceptable for protection by the Mine Safety and Health Administration (MSHA) and the National Institute for Occupational Safety and Health (NIOSH) under the provisions of 30 CFR Part 11.

The employer shall provide a powered, air-purifying respirator in lieu of any negative-pressure respirator specified in Table 3 whenever an employee chooses to use this type of respirator and when this respirator will provide adequate protection to the employee.

Respirator Fit Testing

The employer shall ensure that the respirator issued to the employee exhibits the least possible facepiece leakage and that the respirator is fitted properly.

Employers shall perform either quantitative or qualitative face fit tests at the time of initial fitting and at least every six months thereafter for each employee wearing a negative-pressure respirator. The qualitative fit tests may be used only for testing the fit of half-mask respirators where they are permitted to be worn. The tests shall be used to select facepieces that provide the required protection as prescribed in Table 3.

Protective Clothing

The employer shall provide and require the use of protective clothing, such as coveralls or similar whole-body clothing, head coverings, gloves, and foot coverings for any employee exposed to airborne concentrations of asbestos.

Laundering. The employer shall ensure that laundering of contaminated clothing is done so as to prevent the release of airborne asbestos, tremolite, anthophyllite, actinolite, or a combination of these minerals in excess of the exposure limit.

Protective Clothing for Removal, Demolition, and Renovation Operations. The competent person shall periodically examine worksuits worn by employees for rips or tears that may occur during performance of work.

When rips or tears are detected while an employee is working within a negative-pressure enclosure, rips and tears shall be immediately mended, or the worksuit shall be immediately replaced.

Hygiene Facilities and Practices

The employer shall provide clean change areas for employees required to work in regulated areas.

Exception: In lieu of the change area requirement, the employer may permit employees engaged in small-scale, short-duration operations, as described in this section, to clean their protective clothing with a portable HEPA-filtered vacuum before such employees leave the area where maintenance was performed.

Whenever food or beverages are consumed at the worksite and employees are exposed to airborne concentrations of asbestos, tremolite, anthophyllite, actinolite, or a combination of these minerals in excess of the permissible exposure limit, the employer shall provide lunch areas in which the airborne concentrations of asbestos, tremolite, anthophyllite, actinolite, or a combination of these minerals, are below the action level.

Requirements for Removal, Demolition, and Renovation Operations

Decontamination Area. Except for small-scale, short-duration operations, the employer shall establish a decontamination area that is adjacent and connected to the regulated area for the decontamination of employees contaminated with asbestos, tremolite, anthophyllite, or actinolite. The decontamination area shall consist of an equipment room, shower area, and clean room in series. The employer shall ensure that employees enter and exit the regulated area through the decontamination area.

(1) Clean room. The clean room shall be equipped with a locker or appropriate storage container for each employee's use.

(2) Shower area. Where feasible, shower facilities shall be provided which comply with 29 CFR 1910.141(d)(3). The showers shall be contiguous both to the equipment room and the clean change room, unless the employer can demonstrate that this location is not feasible. Where the employer can demonstrate that it is not feasible to locate the shower between the equipment room and the clean change room, the employer shall ensure that employees:

- remove asbestos, tremolite, anthophyllite, or actinolite contamination from their worksuits using a HEPA vacuum before proceeding to a shower that is not contiguous to the work area; or
- remove their contaminated worksuits, don clean worksuits, and proceed to a shower that is not contiguous to the work area

(3) Equipment room. The equipment room shall be supplied with impermeable, labeled bags and containers for the containment and disposal of contaminated protective clothing and equipment.

(4) Decontamination area entry procedures. The employer shall ensure that employees:

- enter the decontamination area through the clean room
- remove and deposit street clothing within a locker provided for their use
- put on protective clothing and respiratory protection before leaving the clean room

Before entering the enclosure, the employer shall ensure that employees pass through the equipment room.

(5) Decontamination area exit procedures. Before leaving the regulated area, the employer shall ensure that employees:

- remove all gross contamination and debris from their protective clothing
- remove their protective clothing in the equipment room and deposit the clothing in labeled impermeable bags or containers
- do not remove their respirators in the equipment room
- shower prior to entering the clean room
- after showering, enter the clean room before changing into street clothes

Communication of Hazards to Employees

Signs. Warning signs that demarcate the regulated area shall be provided and displayed at each location where airborne concentrations of asbestos, tremolite, anthophyllite, actinolite, or a combination of these minerals may be in excess of the exposure limit. Signs shall be posted at such a distance from such a location that an employee may read the signs and take necessary protective steps before entering the area marked by the signs.

The warning signs shall bear the following information:

DANGER
ASBESTOS
CANCER AND LUNG DISEASE HAZARD
AUTHORIZED PERSONNEL ONLY
RESPIRATORS AND PROTECTIVE CLOTHING
ARE REQUIRED IN THIS AREA

Where minerals in the regulated area are only tremolite, anthophyllite, or actinolite, the employer may replace the term "asbestos" with the appropriate mineral name.

Labels. Labels shall be affixed to all products containing asbestos, tremolite, anthophyllite, or actinolite, and to all containers containing such products, including waste containers. Where feasible, installed asbestos, tremolite, anthophyllite, or actinolite products shall contain a visible label. Labels shall be printed in large, bold letters on a contrasting background. Labels shall be used in accordance with the requirements of 29 CFR 1910.1200(f) of OSHA's Hazard Communication Standard and shall contain the following information:

DANGER
CONTAINS ASBESTOS FIBERS
AVOID CREATING DUST
CANCER AND LUNG DISEASE HAZARD

Where minerals to be labeled are only tremolite, anthophyllite, and actinolite, the employer may replace the term "asbestos" with the appropriate mineral name.

Employee Information and Training. The employer shall institute a training program for all employees exposed to airborne concentrations of asbestos, tremolite, anthophyllite, actinolite, or a combination of these minerals in excess of the action level, and shall ensure their participation in the program.

Training shall be provided prior to or at the time of initial assignment, unless the employee has received equivalent training within the previous 12 months and at least annually thereafter. The training program shall be conducted in a manner that the employee is able to understand. The employer shall ensure that each employee is informed of the following:

- methods of recognizing asbestos, tremolite, anthophyllite, and actinolite
- the health effects associated with asbestos, tremolite, anthophyllite, or actinolite exposure
- the relationship between smoking and asbestos, tremolite, anthophyllite, and actinolite in producing lung cancer
- the nature of operations that could result in exposure to asbestos, tremolite, anthophyllite, or actinolite; the importance of necessary protective controls to minimize exposure including, as applicable, engineering controls, work practices, respirators, housekeeping procedures, hygiene facilities, protective clothing, decontamination procedures, emergency procedures, and waste disposal procedures; and any necessary instruction in the use of these controls and procedures
- the purpose, proper use, fitting instructions, and limitations of respirators as required by 29 CFR 1910.134
- the appropriate work practices for performing the asbestos, tremolite, anthophyllite, or actinolite job and medical surveillance program requirements

Housekeeping

Vacuuming. Where vacuuming methods are selected, HEPA-filtered vacuuming equipment must be used. The equipment shall be used and emptied in a manner that minimizes the reentry of asbestos, tremolite, anthophyllite, or actinolite into the workplace.

Waste Disposal. Asbestos waste, scrap, debris, bags, containers, equipment, and contaminated clothing consigned for disposal shall be collected and disposed of in sealed, labeled, impermeable bags or other closed, labeled, impermeable containers.

Medical Surveillance

The employer shall institute a medical surveillance program for all employees engaged in work involving levels of asbestos, tremolite, anthophyllite, actinolite, or a combination of these minerals at or above the action level for 30 or more days per year, or who are required by this section to wear negative-pressure respirators.

(1) Examination by a physician. The employer shall ensure that all medical examinations and procedures are performed by or under the supervision of a licensed physician and are provided at no cost to the employee and at a reasonable time and place.

(2) Medical examinations and consultations

(a) Frequency. The employer shall make available medical examinations and consultations to each employee covered under these rules on the following schedules:

- Prior to assignment of the employee to an area where negative-pressure respirators are worn
- When the employee is assigned to an area where exposure to asbestos, tremolite, anthophyllite, actinolite, or a combination of these minerals may be at or above the action level for 30 or more days per year, a medical examination must be given within 10 working days following the thirtieth day of exposure. (Exception: No medical examination is required of any employee if adequate records show that the employee has been examined in accordance with this paragraph within the past one-year period.)
- At least annually thereafter
- If the examining physician determines that any of the examinations should be provided more frequently

than specified the employer shall provide such examinations to affected employees at the frequencies specified by the physician

(b) Information provided to the physician. The employer shall provide the following information to the examining physician:

- a description of the affected employee's duties as they relate to the employee's exposure
- the employee's representative exposure level or anticipated exposure level
- a description of any personal protective and respiratory equipment used or to be used
- information from previous medical examinations of the affected employee that is not otherwise available to the examining physician

(c) Physician's written opinion. The employer shall obtain a written opinion from the examining physician.

Recordkeeping

Objective data for exempted operations:

(1) Where the employer has relied on objective data that demonstrate that products made from or containing asbestos, tremolite, anthophyllite, or actinolite are not capable of releasing fibers of asbestos, tremolite, anthophyllite, or actinolite, or a combination of these minerals, in concentrations at or above the action level under the expected conditions of processing, use, or handling to exempt such operations from the initial monitoring requirements, the employer shall establish and maintain an accurate record of objective data reasonably relied upon in support of the exemption.

The record shall include at least the following information:

- the product qualifying for exemption
- the source of the objective data
- the testing protocol, results of testing, and/or analysis of the material for the release of asbestos, tremolite, anthophyllite, or actinolite
- a description of the operation exempted and how the data support the exemption
- other data relevant to the operations, materials, processing, or employee exposures covered by the exemption

The employer shall maintain this record for the duration of the employer's reliance upon such objective data.

(2) Exposure measurements. The employer shall keep an accurate record of all measurements taken to monitor employee exposure to asbestos, tremolite, anthophyllite, or actinolite as prescribed in this section. Note: The employer may utilize the services of competent organizations such as industry trade associations and employee associations to maintain the records required by this section

This record shall include at least the following information: the date of measurement and the operation involving exposure to asbestos, tremolite, anthophyllite, or actinolite that is being monitored.

(3) Training records. The employer shall maintain all employee training records for one year beyond the last date of employment by that employer.

(4) Qualitative and quantitative fit testing procedures. There are several fit test protocols accepted by OSHA. The following is one qualitative method often used for asbestos workers:

Isoamyl Acetate Protocol (Odor Threshold Screening). Three 1-L glass jars with metal lids (e.g., Mason or Bell jars) are required. Odor-free water (e.g., distilled or spring water) at approximately 25°C shall be used for the solutions.

The isoamyl acetate (IAA, also known as isopentyl acetate) stock solution is prepared by adding 1 cm^3 of pure IAA to 800

cm^3 of odor-free water in a 1-L jar and shaking for 30 sec. This solution shall be prepared new at least weekly.

The odor test solution is prepared in a second jar by placing 0.4 cm^3 of the stock solution into 500 cm^3 of odor-free water using a clean dropper or pipette. Shake for 30 sec and allow to stand for 2–3 min so that the IAA concentration above the liquid may reach equilibrium. This solution may be used for only one day.

A test blank is prepared in a third jar by adding 500 cm^3 of odor-free water.

The odor test and test blank jars shall be labeled 1 and 2 for jar identification. If the labels are put on the lids, they can be periodically peeled, dried off, and switched to maintain the integrity of the test.

The following instructions shall be typed on a card and placed on the table in front of the two test jars (i.e., 1 and 2): "The purpose of this test is to determine if you can smell banana oil at a low concentration. The two bottles in front of you contain water. One of these bottles also contains a small amount of banana oil. Be sure the covers are on tight, then shake each bottle for two seconds. Unscrew the lid of each bottle, one at a time, and sniff at the mouth of the bottle. Indicate to the test conductor which bottle contains banana oil."

The screening test shall be conducted in a room separate from the room used for actual fit testing. The two rooms shall be well ventilated but shall not be connected to the same recirculating ventilation system.

The mixtures used in the IAA odor detection test shall be prepared in an area separate from where the test is performed.

If the test subject is unable to correctly identify the jar containing the odor test solution, the IAA qualitative fit test may not be used. If the test subject correctly identifies the jar containing the odor test solution, the test subject may proceed to respirator selection and fit testing.

Respirator Selection. The test subject shall be allowed to pick the most comfortable respirator from a selection including respirators of various sizes from different manufacturers. The selection shall include at least five sizes of elastomeric half facepieces from at least two manufacturers.

The selection process shall be conducted in a room separate from the fit-test chamber to prevent odor fatigue. Prior to the selection process, the test subject shall be shown how to put on a respirator, how it should be positioned on the face, how to set strap tension and how to determine a "comfortable" respirator. A mirror shall be available to assist the subject in evaluating the fit and positioning of the respirator. This instruction may not constitute the subject's formal training or respirator use as it is only a review. The test subject should understand that he or she is being asked to select the respirator which provides the most comfortable fit. Each respirator represents a different size and shape which, if fitted and used properly, will provide adequate protection.

The test subject holds each facepiece up to the face and eliminates those which obviously do not give a comfortable fit. Normally, selection will begin with a half-mask and if a good fit cannot be found, the subject will be asked to test the full facepiece respirators. (A small percentage of users will not be able to wear any half-mask.)

The more comfortable facepieces are noted; the most comfortable mask is donned and worn at least five minutes to assess comfort. All donning and adjustments of the facepiece shall be performed by the test subject without assistance from the test conductor or other person.

If the test subject is not familiar with using a particular respirator, he or she shall be directed to don the mask several times and to adjust the straps each time to become adept at setting proper tension on the straps.

Assessment of comfort shall include reviewing the following points with the test subject and allowing the test subject adequate time to determine the comfort of the respirator:

- positioning of mask on nose
- room for eye protection
- room to talk
- positioning mask on face and cheeks

The following criteria shall be used to help determine the adequacy of the respirator fit:

- chin properly placed
- strap tension
- fit across nose bridge
- distance from nose to chin
- tendency to slip
- self-observation in mirror

The test subject shall conduct the conventional negative- and positive-pressure fit checks (e.g., see ANSI Z88.2-1980). Before conducting the negative- or positive-pressure test the subject shall be told to "seat" the mask by rapidly moving the head from side to side and up and down while taking a few deep breaths. The test subject is now ready for fit testing.

After passing the fit test the test subject shall be questioned again regarding the comfort of the respirator. If it has become uncomfortable another model or respirator shall be tried. The employee shall be given the opportunity to select a different facepiece and be retested if the chosen facepiece becomes increasingly uncomfortable at any time.

Fit Test. The fit test chamber shall be similar to a clear 55-gal drum liner suspended inverted over a 2-ft-diameter frame, so that the top of the chamber is about 6 in. above the test subject's head. The inside top center of the chamber shall have a small hook attached.

Each respirator used for the fitting and fit testing shall be equipped with organic vapor cartridges or offer protection against organic vapors. The cartridges or masks shall be changed at least weekly.

After selecting, donning, and properly adjusting a respirator, the test subject shall wear it to the fit testing room. This room shall be separate from the room used for odor threshold screening and respirator selection and shall be well ventilated as by an exhaust fan or lab hood to prevent general room contamination.

A copy of the following test exercises and rainbow passage shall be taped to the inside of the test chamber:

Test Exercises

- Breathe normally.
- Breathe deeply. Be certain breaths are deep and regular.
- Turn head all the way from one side to the other. Inhale on each side. Be certain movement is complete. Do not bump the respirator against the shoulders.
- Nod head up and down. Inhale when head is in the full up position (looking toward ceiling). Be certain motions are complete and made about every second. Do not bump the respirator on the chest.
- Talk aloud and slowly for several minutes. The following paragraph is called the Rainbow Passage. Reading it will result in a wide range of facial movements and thus be useful to satisfy this requirement. Alternative passages which serve the same purpose may also be used.
- Rainbow Passage: When the sunlight strikes raindrops in the air they act like a prism and form a rainbow. The rainbow is a division of white light into many beautiful colors. These take the shape of a long round arch with its path high above and its two ends apparently beyond the horizon. There is, according to legend, a boiling pot of gold at one end. People look but no one ever finds it. When a man looks for something beyond reach his friends say he is looking for the pot of gold at the end of the rainbow.
- Jog in place.
- Breathe normally.

Each test subject shall wear the respirator for at least 10 min before starting the fit test.

Upon entering the test chamber the test subject shall be given a 6 in. by 5 in. piece of paper towel or other porous absorbent single-ply material, folded in half and wetted with three-quarters of one cm^3 of pure IAA. The test subject shall hang the wet towel on the hook at the top of the chamber.

Allow two minutes for the IAA test concentration to be reached before starting the fit test exercises. This would be an appropriate time to talk with the test subject to explain the fit test, the importance of cooperation, and the purpose for the head exercises, or to demonstrate some of the exercises.

Each exercise described above shall be performed for at least one minute. If at any time during the test the subject detects the banana-like odor of IAA, the test has failed. The subject shall quickly exit from the test chamber and leave the test area to avoid olfactory fatigue. If the test is failed, the subject shall return to the selection room and remove the respirator, repeat the odor sensitivity test, select and put on another respirator, return to the test chamber, and again begin the procedure.

The process continues until a respirator that fits well has been found. Should the odor sensitivity test be failed, the subject shall wait about 5 min before retesting. Odor sensitivity will usually have returned by this time.

If an employee cannot pass the fit test described above wearing a half-mask respirator from the available selection, full facepiece models must be used. When a respirator is found that passes the test, the subject breaks the face seal and takes a breath before exiting the chamber. This is to assure that the reason the test subject is not smelling the IAA is the good fit of the respirator facepiece seal and not olfactory fatigue.

When the test subject leaves the chamber the subject shall remove the saturated towel and return it to the person conducting the test. To keep the area from becoming contaminated the used towels shall be kept in a self-sealing bag so there is no significant IAA concentration buildup in the test chamber during subsequent tests. At least two facepieces shall be selected for the IAA test protocol. The test subject shall be given the opportunity to wear them for one week to choose the one which is more comfortable to wear.

Persons who have successfully passed this fit test with a half-mask respirator may be assigned the use of the test respirator in atmospheres with up to 10 times the PEL of airborne asbestos. In atmospheres greater than 10 times and less than 100 times the PEL (up to 100 ppm), the subject must pass the IAA test using a full-face negative-pressure respirator. (The concentration of the IAA inside the test chamber must be increased by ten times for qualitative fit testing of the full facepiece.)

The test shall not be conducted if there is any hair growth between the skin and the facepiece sealing surface. If hair growth or apparel interferes with a satisfactory fit, then it shall be altered or removed so as to eliminate interference and allow a satisfactory fit. If a satisfactory fit is still not attained, the test subject must use a positive-pressure respirator such as a powered air-purifying respirator, supplied air respirator, or self-contained breathing apparatus.

If a test subject exhibits difficulty in breathing during the tests, she or he shall be referred to a physician trained in respirator diseases or pulmonary medicine to determine whether the test subject can wear a respirator while performing her or his duties.

Qualitative fit testing shall be repeated every six months, at the least.

10
Cost Estimating

Cost estimating[21] is among the most difficult and confusing areas associated with abatement work. Other areas—preparation, safety, clean-up, etc., are all regulated but estimating is still more art than science.

There are several methods used by contractors to arrive at a bid price. The "take-off" method, shown in Table 4, may be the easiest to understand, although it is not quick and simple. To illustrate this method, an example of a removal of ACM from a small school follows.

The building area includes a boiler room, pipe in two classrooms, a 0.25-ft pipe tunnel under a stage in the gym, one 15 ft × 15 ft Transite board wall.

The first step is to carefully review the specification. Some engineers may call out the removal method, order of removal, special schedules, etc. A base bid should reflect the lowest-cost work method which meets the conditions of the specification. It is poor practice to work without a specification and detailed contract. Most contractor-owner disputes arise when there is no clear document describing the work. For example: is the contractor to provide his own water; is he to reinsulate or replace walls; is he to encapsulate at the wall or remove an entire wall to get to a few feet of ACM? Who determines the job is done? In a recent

Table 4. The Take-Off Method of Cost Estimating

Item	Cost
Prework walk-through including travel	$ 400
Scaffold rental	350
Wood	200
Poly filters & encapsulant	175
Tape	55
Glove bags	225
Disposal bags	115
Work crew—4 men, 10 days—direct cost	1,600
Benefits	500
Worker medicals	300
Tyvek suits & resp cartridges	370
Subcontractor to replace wall (bid)	2,100
Truck rental	300
Neg air machine filters	250
Vacuum filters	75
Sample pump rental	400
Sample analysis	1,100
Cleanup crew	800
Hotel costs	450
Reimbursable meals	500
Industrial hygienist/PE	1,000
Shower	300
Tools (disposable knives, etc.)	250
Disposal (landfill charges)	225
Reinsulate pipes (subcontractor)	800
Reinsulate boiler (subcontractor)	1,800
Est. equipment repairs	150
Meeting when job is done	200
Subtotal	$14,990
Overhead	110
Profit	1,800
Insurance	3,000
Contingency	1,000
	$20,900

unfortunate abatement program an experienced contractor took on removal and refireproofing. After the job was completed, the owner said he expected 3 in., not 2 in., of fireproofing, and he expected the contractor to pay for certification of the fireproofing and for any loss due to delays in getting the certification. Then he brought in an "expert" who looked at surfaces after the fireproofing was applied and said they were not clean enough. Of course during the project the schedule was the most important concern but now the owner is unhappy, the contractor still has not been paid, and some attorneys on both sides are charging by the hour.

The take-off method of cost estimating requires that every cost associated with the project be identified.

Each line item is based on further detailed take-offs of information. For example, the amount of tape is based on the number of seams and amount of wall to be taped. The subcontractor amounts are based on actual written estimates of bids by the subcontractor. To establish the need for 4 men, 10 days, a work plan is required. For example:

1st day	1 person check inventory and order wood, plastic, safety gear, etc., to be delivered by next Monday.
	Assume 3 man-days to get all the material to the site.
7th day	1 foreman and 2 workers set up scaffold, install containment around boiler.
8th & 9th days	1 foreman, 2 workers on boiler, 1 man bagging.
10th day	1 foreman, 2 men scrub down boiler, 1 man to run vac and materials.
11th day	1 foreman, 2 men remove containment cleanup.
12th day	1 man on-site, set up tunnel containment.

13th & 14th days	1 foreman, 3 workers in tunnel—glove bag work goes slow because of available space and problems with lights.
15th day	1 foreman and 1 worker clean up tunnel.
16th & 17th days	1 foreman and 2 workers contain and seal off 2 classrooms and remove wall.
18th day	1 foreman and 2 workers remove radiation ACM.
19th & 20th days	1 foreman and 2-man crew glove bag pipe insulation, 6 ft/hr because suspended ceiling is in the way.
21st & 22nd days	1 foreman and 1 worker replace ceiling tiles and pack up materials.
23rd day	1 man takes ACM to landfill; returns rental truck; 4 hr
24th day	1 man cleans equipment and stores.
25th day	1 foreman, 1 worker clean up after subcontractors.
Total	400 hr

The overhead $110 is to cover secretarial time in payroll, paying bills, etc. There are no absolute methods to ensure an accurate estimate. An astute contractor may propose sealing off the tunnel, a one-man-day job, to lower costs. Check every line item to be certain it is needed and the owner is aware of exactly what is to be provided. Replacing the ceiling, moving desks, cleaning up after subcontractors, and dozens of other little items can easily wipe out any profit if not accounted for. They can also cause an inflated estimate and awarding of the contract to someone else. The most successful contractors look for legal shortcuts. For example, if the boiler were to be replaced the abatement contractor might remove it as one piece! Super-sucker trucks which have HEPA filters and 8-in. suction hoses can make a two-week cleanup into a two-day job. Perhaps the replacement of the wall ($2000) could be done with gunite for $600, or perhaps the reinsulation could be done by the asbestos abatement contractor under the same foreman while removal is going on. Typical equipment costs, based on 1987 catalog prices, can be found in Appendix D, to assist in estimating costs. Attention to every detail is the hallmark of a successful and profitable removal team.

11
Legal Aspects and Insurance

LEGAL ISSUES

There are three legal areas usually of interest in asbestos abatement: of most immediate concern is contract law; secondly, the rules and regulations of EPA and OSHA; and finally, the area of tort law, which we commonly think of in terms of being sued for damages.

Contract law is based on custom, on state or federal law, and to a large extent on the Uniform Commercial Code (UCC). In the vast majority of cases where disputes arise, no proper written contract has been executed. By "proper written contract" we mean one which spells out what will be done or where the work stops, what payment criteria are used, when payment is due, and even details such as who pays for power and water, who provides toilets, and especially who decides when the job is finalized and what constitutes the criteria.

Although contract law is complex and there are apparent exceptions to most rules, some basic understanding is necessary for anyone entering into a business agreement.

A contract is an agreement in which two persons promise to exchange something of value. One well-known legal guideline is the restatement of contracts. It states that a contract is a promise or set of promises for breach of which the law gives a remedy, or which the law recognizes as a duty.

Three categories of contract exist. (1) An expressed contract is formed by spoken or written words which are understood by both parties to create a duty. (2) Implied contracts are formed in other ways. For example, if a contractor has been cleaning the sidewalk for the last two years and has been paid by the owner $10 each time, then when the owner calls and says, "It's time to clean my walk again," it is implied that the owner intends to pay $10 if the walk is cleaned. (3) The third category is quasi contracts in which the law provides for payment to prevent unjust enrichment. An example of this type is a contractor who has started work and finds that a water pipe must be replaced. He contacts the owner and begins discussing the cost. No agreement is reached but the foreman believes the owner authorized replacement and installs a new pipe, while the owner watched it being replaced and said nothing. In this case the contractor can probably get the fair market value of the replacement, or at least his cost, as compensation from the owner.

A contract which is proper in all important aspects is called valid or completely enforceable. A contract to commit a crime is termed void and cannot be enforced. A contractor who agrees to remove asbestos in an illegal manner may or may not have a void contract. A contract obtained under false pretenses or fraud may also be unenforceable.

Some basic technical requirements must be met in all valid contracts. First there must be mutual assent. This means that each person must agree to and understand the contract. There must be an offer and acceptance. Taking all the circumstances into consideration, if one person offers something of value such as money or a promise to do work and the other agrees to do the work or pay for having it done, then there is mutual assent. One area where problems arise is in preliminary negotiations or advertisements. Prior to a contract, usually there is some fact exchange, bargaining, or other preliminary activity. Oral statements such as "what if" or "make me an offer" are not usually contracts or agreements. The words must say or clearly imply offer and acceptance of the offer. A conditional acceptance, such as "I'll do it if you do the cleanup," is really a counteroffer and has no binding effect until both parties agree. It is the court, not the parties, that finally determines when an agreement was made, if there is a dispute. An agreement must be definite in its terms. An agreement to remove some asbestos for some payment is vague; an agreement to provide six men and required materials to remove asbestos for two weeks for $15,000 is definite and implies the workers will follow industry standards and provide reasonable output every day.

The courts will "supply" reasonable terms including price, time, and quantity when the court (the judge) deems it appropriate and fair. While the courts will sometimes provide reasonable terms for a contract, they will not ensure satisfaction with the work or a profit for the contractor.

One problem which comes up often is the stoppage of work. If the contractor walks off the job, must the owner pay? If the owner stops work or hires someone else to finish, must the contract be paid? The answers are not simple. If a contractor is unable to finish, or leaves over a dispute, the owner can recover the added cost of having the work finished, but usually must pay the remainder of the original contract to the contractor. If the owner stops work, he may owe the contractor for the work completed plus other expenses and possibly even profit. Third-party beneficiaries can become another area of concern. To illustrate this, take a case where the federal government contracts to remove asbestos from a local school leased to the state school system by a commercial firm. If the work is not completed, the children (through their parents) may have a right, the school system and even the building owner may all have a right, to enforce the contract or sue for damages. Although every case is different, many asbestos removal projects affect many people in addition to the contractor and the person who pays the bill. Our legal system is complex, and in many cases some of the affected persons who were not even party to the contract may have enforceable rights.

Only when the contract is in writing and clearly states cost, basis for payment, type of abatement, extent of abatement, criteria for completion, times, dates, and every other detail, is it likely that everyone will be satisfied and lawsuits avoided.

Tort law is also based on customs and statute. In its basic form, when an existing duty is breached, and damages occur as a result of the breach, a tort exists. Most of the high-dollar lawsuits are made under tort law.

A contractor has a duty to protect everyone from exposure to asbestos dust which results in any way from his work. This duty is implied because asbestos is known to be dangerous. People are not expected to protect themselves and the contractor has control of the asbestos; therefore, the law imposes a "duty to protect" on the contractor. The contractor's workers may be prevented from bringing a lawsuit by workmen's compensation laws, but the worker's family might not be barred from taking action. Schoolchildren, neighbors, visitors, and even a thief breaking into the building, must be protected.

Simply complying with OSHA and EPA rules does not ensure the duty has not been breached. For example, in a school with lower grades, the warning signs may not be judged enough to keep small children out of asbestos-waste containers. A wind may knock down a containment and expose neighbors to asbestos. Probably the contractor, the school system, and even the state inspector can all be found guilty of breaching some duty. The possible future cancer, the worry about cancer (emotional harm), loss of property value because of possible contamination, cleanup costs, etc., are all causes for legal action. In addition, if the breach of duty is found in any way intentional or grossly negligent, or for other reasons, punitive damages may be awarded. Not only is the contractor (and school as the employer/owner) responsible for cleaning up the neighborhood, but residents may have a few million dollar's worth of worry, homes may drop in value, a real estate agent may lose a sale, and dozens of others may find losses they blame on the incident. The court may find that asbestos removal is a known dangerous activity, like making fireworks, and even though the contractor complied with all regulations and was careful, he/she still must pay for all damages. If it is determined by the court that the contractor could have done more (for example, kept a 24-hour team on site) then perhaps a few million dollars of punitive damages could be assessed.

The best advice the author has to offer is to comply with all rules and standards, then review all operations and document every safety measure taken. Since the building owner/operator also has some responsibilities, everyone should be involved in decisions. A contract proposal may state that for an added $2,000, a reinforced plastic may be used for outdoor containments; however, it is not required by OSHA. The decision not to spend the extra money must be weighed against the total risk. It is entirely possible to take a legal $40,000 project and spend over $100,000 by being careful and safe, but still be sued. The best position for an owner or contractor is one in which all standards are met, all current accepted work practices are followed, and all persons who could be exposed to asbestos are protected. In practical terms it may be impossible to meet realistic budgets while spending dollars on every possible safety concern. An experienced removal team which includes engineers, industrial hygienists, safety professionals, and legal experts to weigh costs and benefits is one approach used on some of the more successful projects.

REGULATORY ISSUES

Most of this text deals with EPA, OSHA, or de facto standards of groups such as the National Asbestos Council. Almost every week another rule or standard is proposed or promulgated. The full-time professional should read the *Federal Register* and the journals, attend two or three conferences a year, and meet often with regulatory officials (state and federal) to keep current. For part-time asbestos managers, there are abstract services covering EPA and OSHA and even a few dedicated to asbestos.

Studies on health effects, removal methods, substitute methods, containment, and every other aspect of asbestos abatement are ongoing throughout the world. Even the most dedicated professional is unlikely to be current on every aspect of the subject.

In view of the constant changes and new information, it is the responsibility of every professional to keep as current as possible.

INSURANCE

Types of Insurance[10]

In the past, errors and omissions liability insurance has been written on an "occurrence" basis. If an incident "occurs" while the policy is in force, coverage is afforded even if the actual claim is made some years later and even if the insured is no longer insured by the same carrier. Occurrence policies can result in great losses to carriers who have not received premiums over a period of time, especially given the long latency periods for asbestos-related diseases. As a result, the carriers have been adding exclusions to existing policies for asbestos-related, third-party claims and generally have changed the coverage from "occurrence" to "claims made."

Under a "claims made" policy, coverage exists if a claim is made while the policy is in force. In certain situations, a claim may be made during an extended ("tail") reporting period. The tail may require an additional premium. For many risks, the difference between occurrence and claims-made coverage is not significant, since the liability-causing event is obvious, and claims are generally asserted shortly after the event occurs. However, the release of asbestos fibers caused by a "planned response action" may not be obvious and injury may not be detected until 20 to 40 years later. Claims-made coverage may not be of value in such cases if (1) the insured changes insurance carriers before a claim is made, or (2) the carrier withdraws from the market before a claim is filed. Nevertheless, it is likely that claims-made coverage will be the type of insurance available in the future and an analysis must be made by the insured as to what coverage is actually being purchased. There is no single definition of what "claims made" or "occurrence" means; thus, it is mandatory that the insured read and understand the coverage afforded under the policy. All exclusions, conditions, and definitions should be carefully analyzed.

There are several important considerations in making an analysis of available insurance coverage or in specifying it:

- True "occurrence" coverage is rare. The terms of the policy must be reviewed carefully. Some "occurrence" policies have conditions or exclusions that negate coverage. The name of the policy makes no difference. Claims-made policies may, in some situations, cover claims which arose in prior years, similar to "occurrence" policies.
- The insurance certificate itself provides little or no information regarding the specifics of coverage. The policy itself must be reviewed.

- The insurance carrier should be carefully evaluated. Does the carrier understand the industry, and is it committed to writing proper coverage? Again, the policy terms are important.

Problems that have developed from this area can be illustrated by the following examples:

- a general liability insurance policy issued for asbestos work which excludes coverage for personal injury attributable to airborne mineral fibers. Of course, asbestos is a mineral fiber and is normally dangerous only when it is airborne, and thereafter inhaled.
- an errors and omissions policy written for a consultant which includes a "pollution exclusion" excluding coverage for any personal injury or property damage caused by a broad list of substances, including asbestos. This policy provides no coverage for asbestos risks.
- a general liability "occurrence" policy that excludes "anticipatory damages," which is defined as damages that are claimed to have been caused by asbestos, but which cannot be proved, due to the fact that the asbestos-related disease has not yet manifested itself. This situation is perhaps the type of claim which can most often be expected, but no coverage is provided in these circumstances.

Asbestos Liability

The massive litigation in which asbestos manufacturers are involved, and in which their insurance carriers have become involved, have led to a "gun-shy" attitude on the part of carriers and reinsurers toward insuring those in the asbestos control industry, despite the difference in the employment and work practices of asbestos manufacturers.

BONDING

The difficulties in obtaining insurance have spread to the bonding industry. Traditionally, two types of bonds have been required in the construction industry to protect the owner or lender against the contractor's financial default:

- payment bonds, under which a surety company agrees to pay for labor and materials supplied to a project in the event the contractor fails to do so; and
- performance bonds, under which a surety agrees to complete performance of a project, if the contractor fails to do so.

Abatement contractors who have had their insurance canceled or not renewed are experiencing difficulties in obtaining bonding. Bonding companies rely on the financial ability of the principal (the contractor) to respond to claims under payment and performance bonds. If a company is not insured against catastrophic liability, the financial underpinnings of the company are weakened, and the bonding company becomes apprehensive over issuing bonds.

There are numerous legal considerations involved in the evaluation of insurance and bonding coverage. The cost of insurance for asbestos abatement is significant, and if such expense is going to be undertaken, the coverage obtained should be satisfactory. While there are no easy solutions in this decisionmaking process, it is mandatory that contractors, consultants, and owners become knowledgeable purchasers of insurance.

12
Asbestos Hazard Emergency Response Act

ASBESTOS IN SCHOOLS RULES

In October 1986 the Asbestos Hazard Emergency Response Act (AHERA) was signed into law. Included in this Act are provisions directing the EPA to establish rules and regulations addressing asbestos-containing materials in schools. The final AHERA regulations (rules) became effective October 17, 1987.

This regulation has far-reaching economic and political impact. It will serve as a model for regulations in commercial, industrial, and even private sector buildings in the future.

SPECIFIC PROVISIONS OF THE LAW

AHERA rules apply to all public and private elementary and secondary schools in the United States and its territories, and to American schools on military bases in foreign countries.

Schools' responsibilities include numerous activities. Fines of up to $25,000 per day and jail terms for violations are included. Many of the major requirements are noted in the following list.

Selected Requirements

1. Designate a person to ensure that AHERA requirements are properly implemented.
2. Inspect and identify friable and nonfriable asbestos-containing material (ACM).
3. Monitor and periodically reinspect.
4. Develop and update management plans.
5. Determine and implement response actions.
6. Develop and implement operations and maintenance programs.
7. Notify parents, building occupants, and outside contractors of ACM identified in the building.
8. Ensure that accredited persons perform these required activities under AHERA.
9. Inspect schools before October 12, 1988, and have accredited inspectors reinspect at least every three years, and schools reinspect every six months. By October 12, 1988, schools must prepare and submit, to an agency designated by the Governor, an asbestos management plan for each building. The plan must be kept up-to-date.

Under the program each school can determine whether suspect material contains asbestos. To accomplish this, bulk sampling of materials must be conducted in the manner specified in the law. Bulk samples are to be analyzed for asbestos by laboratories accredited by the National Bureau of Standards, or laboratories with accreditation from EPA.

All friable asbestos-containing building material (ACBM) and assumed ACBM must be located and categorized as to present condition, potential for damage, and type of material. Nonfriable ACBM and assumed ACBM must be identified and documented, but not assessed.

Some particular conditions require specified response actions. For example, any building where friable ACBM is present, or assumed to be present, must develop and implement an operations and maintenance (O&M) program. The O&M program must provide for surveillance of ACM at least every six months. The plan is required to contain information specified in AHERA, and schools must begin implementation of the management plan by July 9, 1989. A management plan must be prepared and submitted for any building to come into service after October 12, 1988, prior to its use as a school.

Recordkeeping is one of the added requirements of the law. A detailed written description of any preventative or response action taken for ACBM must be appended to the management plan. Records of air monitoring, training, surveillance, cleaning, O&M, fiber release episodes, and reinspections must be maintained and added to the management plan.

Warnings to affected persons are another requirement. Warnings must be posted adjacent to any ACBM located in maintenance areas of a building. Warning labels must read:

CAUTION:
ASBESTOS. HAZARDOUS.
DO NOT DISTURB WITHOUT
PROPER TRAINING AND EQUIPMENT.

Training of individuals dealing with the asbestos problem is regulated under the law. Added information on this topic can be found in Appendix D; however, the following summary will explain most of the requirements.

Nature of the required training under AHERA:

- Building Inspector—three-day course with field work and exam; half-day annual refresher training
- Management Planner—three-day building inspectors' course, plus two additional days and exam; one-day annual refresher training
- Project Designer—three-day course with field work and exam or abatement supervisor course; one-day annual refresher training

- Abatement Supervisor—four-day course with hands-on training and exam; one-day annual refresher training
- Abatement Worker—three-day course with hands-on training and exam; one-day annual refresher training

STATE AND LOCAL REGULATIONS

Each state is to adopt an accreditation plan at least as stringent as the EPA model, and an agency of the state is to be named to receive and review the local education agency's (LEA) management plan.

Several provisions in AHERA encourage states to develop their own regulatory programs. For example, states are encouraged to establish and operate training and certification programs for the various categories of asbestos professionals, as long as the programs are at least as stringent as AHERA's Model Plan. In addition, some states have established requirements that exceed EPA's in the area of notification of abatement disposal of asbestos-contaminated waste. Building Inspectors and Management Planners should consult state and local regulatory agencies in their areas.

Requirements and Exclusions Under AHERA

The AHERA Rule requires that all suspect materials be identified, located, and documented; and that friable suspect materials be assessed and classified.

Under certain circumstances, the local education agency may not be required to inspect their buildings. The criteria for exclusion are:

1. An accredited inspector has determined that friable asbestos-containing building material (ACBM) was identified during an inspection conducted prior to October 17, 1987.

2. An accredited inspector has determined that nonfriable ACBM was identified during an inspection conducted prior to October 17, 1987.

3. An accredited inspector has determined (based on sampling and inspection records) that no ACBM is present and the records show that the area was sampled before October 17, 1987.

4. The appropriate state agency has determined that no ACBM is present and the records show that the area was sampled before October 17, 1987.

5. The accredited inspector has determined (based on inspection and sampling records conducted before October 17, 1987) that suspected ACBM will be assumed to be ACM.

6. The accredited inspector has determined that no ACBM is present where asbestos removal operations have been conducted before October 17, 1987.

7. An architect or project engineer responsible for construction of a new school building built after October 12, 1988, or an accredited inspector, signs a statement that no ACBM was specified as a building material, and to the best of his/her knowledge, no ACBM was used as a building material.

Also, if ACBM is subsequently found to be present, the LEA will have 180 days to comply with the AHERA inspection requirements.

Key People in AHERA Rules

The Asbestos Hazard Emergency Response Act suggests certain education prerequisites for Building Inspectors and Management Planners. However, states are free to adopt standards which may be higher or lower than the federal suggestions.

Functions of the AHERA Professionals

The Building Inspector is the person responsible for (1) determining whether ACM is present in a building and (2) assessing physical characteristics of the ACM and of the building. The Management Planner then uses this information to estimate the degree of current or potential hazard posed by the ACM, and to develop a plan for managing the ACM.

The selection of a response action should be based upon a number of evaluating factors, including (1) hazard assessment, (2) costs—initial and long-term, and (3) life of the facility.

The Management Planner determines which response action is appropriate for all ACBM identified in the building. The single most important factor in determining a response action must be the health and safety of the building occupants. Once this factor has been gauged, all other factors should be incorporated into the final decision. In so doing, the Planner will find it advantageous to consult with other professionals.

SUMMARY OF INSPECTION REPORT AND MANAGEMENT PLAN FOR AHERA

The AHERA Rule requires that the following key items of information be included in the Inspection Report:

- a list of identified homogeneous areas classified by type of material (surfacing material, thermal system insulation, or miscellaneous material)
- the location of homogeneous sampling areas and individual sampling locations, the location of friable suspected material assumed to be ACBM, the location of nonfriable suspected material assumed to be ACBM, and the dates of sampling
- approximate square or linear footage of any homogeneous or sampling area where material was sampled for ACM
- a copy of the laboratory analyses for each bulk sample and designation of each homogeneous area as ACM or non-ACM, including the dates of sample analyses

- the physical assessment of ACBM and suspect ACBM and placement into one of the following categories:

    1. damaged or significantly damaged thermal system insulation ACBM
    2. damaged friable surfacing ACBM
    3. significantly damaged friable surfacing ACBM
    4. damaged or significantly damaged friable miscellaneous ACBM
    5. ACBM with potential for damage
    6. ACBM with potential for significant damage
    7. any remaining friable ACBM or friable suspect ACBM

- the name and signature of each accredited inspector collecting samples, the state of accreditation, and, if applicable, his or her accreditation number

According to AHERA, the following key elements comprise the Management Plan:

- general building description and a summary of the Inspection Report
- descriptions of hazard assessments for all ACBM and all suspect material assumed to be ACBM
- recommended preventative measures (operations and management program) and/or response actions for any friable ACBM
- location where preventative measures and response actions are to be implemented
- reasons for selecting the measures and actions
- schedule for implementation
- identification of ACBM which remains after response actions are taken

- plan for periodically reinspecting ACBM
- program for informing workers and building occupants
- evaluations of resources needed to implement the management plan

CLASSIFICATION OF ACBM IN SCHOOLS

The AHERA Rule requires that the Building Inspector's Report include a classification of all friable ACBM into one of the following categories:

- significantly damaged or damaged thermal system insulation (TSI)
- significantly damaged friable surfacing material
- damaged friable surfacing material
- significantly damaged or damaged friable miscellaneous materials
- friable ACM with a potential for damage
- friable ACBM with a potential for significant damage
- any other friable ACBM

HAZARD ASSESSMENT

The hazard assessment is conducted by combining the level of potential disturbance with the current condition of the ACM to indicate overall hazard potential.

The rankings of potential hazard range from 7 = most hazardous to 1 = least hazardous. The highest rank is reserved for ACM in poor condition.

AHERA definitions for friable surfacing and miscellaneous materials:

Significantly Damaged—ACBM where the damage or deterioration is extensive and severe.

Damaged—ACBM where the damage or deterioration is characterized by inadequate cohesion or adhesion.

AHERA definitions for thermal system insulation:

Significant Damage or Damage—insulation which has lost its structural integrity or its covering, or is crushed, waterstained, gouged, punctured, missing, or not intact.

Factors to be Used in Determining the Potential for Disturbing Potential ACBM

Potential for Contact with the Material

High: (1) service workers work in the vicinity of the material more than once per week, or (2) the material is in a public area (e.g., hallway, corridor, auditorium) and accessible to building occupants.

Moderate: (1) service workers work in the vicinity of the material once per month to once per week, or (2) the material is in a room or office and accessible to the occupants.

Low: (1) service workers work in the vicinity of the material less than once per month, or (2) the material is visible but not within reach of building occupants.

Influence of Vibration

High: (1) loud motors or engines present (e.g., some fan rooms), or (2) intrusive noises or easily sensed vibrations (e.g., major airports, a major highway).

Moderate: (1) motors or engines present but not obtrusive (e.g., ducts vibrating but no fan in the area), or (2) occasional loud sounds (e.g., a music room).

Low/None: none of the above.

Potential for Air Erosion

High: high-velocity air (e.g., elevator shaft, fan room).

Moderate: noticeable movement of air (e.g., air shaft, ventilator air stream).

Low/None: none of the above.

The ranking can be put into a table form as follows:

Classification for Hazard Potential

Hazard Rank	ACM Condition	ACM Disturbance Potential
7	Poor	Any
6	Fair	High
5	Fair	Moderate
4	Fair	Low
3	Good	High
2	Good	Moderate
1	Good	Low

RECORDKEEPING UNDER AHERA

In general, the recordkeeping system must track three types of data: data on the physical condition of the ACBM, actions taken on the ACBM, and the data associated with the personnel involved with the asbestos management program.

The required recordkeeping for personnel includes the identity, training, medical monitoring, and exposure of persons. This information should be recorded in a form which will be available for a period of at least 30 years, which is also required by OSHA.

The various types of documents and records to be included in the recordkeeping system are outlined below:

1. For each preventive measure or response action taken:

- detailed written description of the measure or action
- methods used
- location
- justification for why a specific measure or action was selected
- start and completion dates of all work
- names and addresses of all contractors involved and accreditation information
- if ACM was removed, name and location of storage or disposal sites

2. For any air sampling conducted:

- name and signature of person collecting samples
- date and location where samples were collected
- name and address of laboratory analyzing samples
- date and method of analysis
- results of analysis
- name and signature of analyst

3. For persons required to be trained for maintenance and repair operations, training records must be maintained:

- employee's name and job title
- date training completed
- location of training and training organization's name
- number of hours of training

4. For each time periodic surveillance is performed:

- inspector's name
- date of the surveillance

- notation of changes (or lack of) in the condition of the ACBM

5. For each time cleaning is performed:

   - name of person(s) doing cleaning
   - date of cleaning
   - locations cleaned
   - methods used in cleaning

6. For each time operations and maintenance activities are performed:

   - name of person(s) performing activities
   - start and completion dates of action
   - locations
   - description of activity, including preventive measures taken
   - if ACBM removed, name and location of storage/disposal site

7. For each time maintenance activities other than small-scale, short-duration activities are undertaken:

   - name, signature, and state of accreditation of each person involved in activity
   - start and completion dates of project
   - location(s)
   - description of project, including preventative measures taken
   - if ACBM removed, name and location of storage/disposal site

8. For each fiber release episode:

   - date of episode

- location
- method of repair
- preventative measures or response action taken
- name(s) of person(s) performing work
- if ACBM is removed, name and location of storage/disposal site

9. Documentation suggested but not required:

    - complete historical blueprint of facility, if available
    - documentation on materials/products used in construction or renovation of the facility that may contain asbestos (include any correspondence with manufacturers)
    - location and photographs of warning signs and barriers placed to prevent unauthorized access to areas of ACBM
    - required state and federal forms dealing with notification and compliance
    - all correspondence pertaining to asbestos in the facility
    - copies of notification statements, press releases, meeting agenda (with attendance rosters)

The reasons for maintaining complete and detailed records of asbestos management are many. Documentation can expedite response actions and make future renovation in any facility easier. The legal liabilities involved with asbestos are another reason to maintain thorough records. The more thorough the documentation, the more defensible the actions taken. Further, poor or sloppy recordkeeping could imply callousness toward employees, building occupants, and the public. In the case of LEAs, records are kept because they are required by AHERA.

Glossary

ABIH. American Board of Industrial Hygiene.

ACM. Asbestos-containing material.

acoustical insulation. The general application of use of asbestos for the control of sound due to its lack of reverberant surfaces.

acoustical tile. A finishing material in a building usually found in the ceiling or walls for the purpose of noise control.

aggressive sampling. Air sampling which takes place after final clean-up while the air is being physically agitated to produce a "worst case" situation.

AIA. Asbestos Information Association.

AIA. American Institute of Architects.

AIA. American Insurance Association.

AIHA. American Industrial Hygiene Association.

AIHA-accredited laboratory. A certification given by the AIHA to an analytical laboratory that has successfully participated in the "Proficiency Analytical Testing" program for quality control as established by the National Institute for Occupational Safety and Health.

airborne asbestos analysis. Determination of the amount of asbestos fibers suspended in a given amount of air.

air diffuser. A device designed to disperse an air stream throughout a given area.

air lock. A system of enclosures consisting of two polyethylene curtained doorways at least three feet apart that does not permit air movement between clean and contaminated areas.

air man. An industrial hygienist or other qualified individual who collects air samples and monitors the asbestos abatement worksite.

air monitoring. The process of measuring the airborne fiber concentration of a specific quantity of air over a given amount of time.

air plenum. Any space used to convey air in a building or structure. The space above a suspended ceiling is often used as an air plenum.

algorithm. A universally accepted procedure developed for the purpose of solving a particular problem. Algorithms developed for asbestos provide a numerical index for evaluating a degree of hazard in a particular area. The Sawyer Algorithm and the Ferris Index are two, but neither are widely used today.

alveolar macrophages. Highly specialized mobile cells in the lungs that attempt to engulf and digest such lung hazards as dusts or fibers.

alveoli. Located in clusters around the respiratory bronchioles of the lungs, this is the area in which true respiration takes place.

ambient air. The surrounding air or atmosphere in a given area under normal conditions.

amended water. Water to which a chemical wetting agent (surfactant) has been added to improve penetration into asbestos-containing materials that are being removed.

amosite. An asbestiform mineral of the amphibole group containing approximately 50% silicon and 40% iron (II) oxide, made up of straight, brittle fibers, light gray to pale brown in color.

amphibole minerals. One of the two major groups from which the asbestiform minerals are derived, distinguished by their chain-like crystal structure and chemical composition.

ANSI. American National Standards Institute.

approved landfill. A site for the disposal of asbestos-containing and other hazardous wastes that has been given EPA approval.

asbestiform minerals. Minerals which, due to their crystal structures and chemical composition, tend to be separated into fibers and can be classified as a form of asbestos.

asbestos. A generic name given to a number of naturally occurring hydrated mineral silicates that possess a unique crystalline structure, are incombustible in air, and are separable into fibers. Asbestos includes the asbestiform varieties of chrysotile (serpentine); crodidolite (riebeckite); amosite (cummingtonite-grunerite); anthophyllite; and actinolite.

asbestos abatement. Procedures to control fiber release from asbestos-containing materials in buildings.

asbestos control. Minimizing the generation of airborne asbestos fibers until a permanent solution is developed.

asbestos exposure assessment system. A decision tool which can be used to determine the extent of the asbestos hazard that exists in a building, and which can also be used to develop corrective actions.

asbestos fibers. Fibers with their length being greater than five μm (length-to-width ratio of 3:1), generated from an asbestos-containing material.

asbestos standard. References to the OSHA requirements in the general industry standards regarding asbestos exposure (29 CFR 1910.1001), and EPA National Emission Standards for Hazardous Air Pollutants (NESHAPS) (40 CFR 61, subpart M).

asbestosis. A nonmalignant, progressive, irreversible lung disease caused by the inhalation of asbestos dust and characterized by diffuse fibrosis.

aspect ratio. The length of a fiber vs. its width.

atmospheres immediately dangerous to life or health. A hazardous atmosphere to which exposure will result in serious injury or death in a matter of minutes, or cause serious delayed effects.

atmosphere-supplying respirators. Respiratory protection devices which exclude workplace air altogether and provide clean air from some independent source.

bid. A statement of the price at which a contractor will complete a given project.

"blue book." EPA publication of March 1983 titled "Guidance for Controlling Friable Asbestos-Containing Materials in Buildings." Now replaced by 1985 revised edition.

bridging encapsulant. The application of a sealant over the surface of asbestos-containing material to prevent the release of asbestos fibers.

bronchi. Primary branches of the trachea (windpipe).

bronchogenic cancer. An abnormal cell growth in the primary branches of the trachea (windpipe).

cancer. A cellular tumor which normally leads to premature death of its host unless controlled.

carbon monoxide. A highly toxic colorless and odorless gas.

ceiling concentration. The maximum allowable level of toxic material that can be present at any given point in time.

cementitious. Asbestos-containing materials that are densely packed, granular, and friable.

CFM. Cubic feet per minute.

chrysotile. The only asbestiform mineral of the (white asbestos) serpentine group which contains approximately 40% each of silica and magnesium oxide. It is the most common form of asbestos used in buildings.

CIH. An industrial hygienist who has been granted certification by the American Board of Industrial Hygiene.

cilia. Tiny hair-like structures in the windpipe and bronchi of the lung passages that help force undesirable particles and liquids up and out of the lungs.

claustrophobia. The fear of being in enclosed or narrow spaces.

clean area. The first stage of the decontamination enclosure system in which workers prepare to enter the work area.

clerk of the works. A person who coordinates and oversees all activities on an asbestos abatement job site.

closed circuit SCBA. A self-contained respiratory protection device in which the air is rebreathed after the exhaled carbon dioxide has been removed and the oxygen content restored.

columns. The building components which support the structural beams.

compressed oxygen. A self-contained respiratory protection cylinder-type closed device in which air is supplied from a circuit SCBA compressed air cylinder. The exhaled air is filtered to remove carbon dioxide, and additional breathing air is provided.

concrete-like asbestos. Hard, nonfriable asbestos-containing material that requires a mechanical force to penetrate its surface.

contaminated items. Any objects that have been exposed to airborne asbestos fibers without being sealed off or isolated.

continuous flow airline device. A respirator that maintains a constant airflow to the wearer.

contract specifications. A set of guidelines that a contractor must follow when conducting an asbestos abatement job.

core certified. A misleading term used by some industrial hygienists in training (see CIH).

CPSC. Consumer Product Safety Commission.

criteria document. NIOSH publications that address toxic materials, analytical methods, personal protective equipment, etc.

decontamination enclosure system. A series of connected rooms with polyethylene curtained doorways for the purpose of preventing contamination of areas adjacent to the work area.

demand airline. A respirator in which air enters the device facepiece only when the wearer breathes in.

dirty area. Any area in which the concentration of airborne asbestos fibers exceeds 0.01 f/cm^3, or where there is visible asbestos residue.

dispersion staining. Used in conjunction with polarized light to identify bulk samples. A particle (fiber) identification technique based on the difference between light dispersion of a particle (fiber) and a liquid medium in which it is immersed.

duct tape. Heavy gauge tape capable of sealing joints or adjacent sheets of polyethylene.

dust mask. Single-use or disposable dust respirator with a low protection factor.

electron microscopy. A method of asbestos sample analysis which utilizes an electron beam to differentiate between fibers.

employee notification. Informing employees or building occupants if asbestos is present in the building; also informing them of the hazards associated with asbestos exposure, what is being done to eliminate the problem, etc.

employer's liability. Legal responsibility imposed on an employer requiring him/her to pay damages to an injured employee.

encapsulant. A substance applied to asbestos-(sealant) containing material with a bonding or sealing agent to prevent the release of airborne fibers.

encapsulation. The coating of asbestos-containing material with a bonding or sealing agent to prevent the release of airborne fibers.

EPA. Environmental Protection Agency.

EPA regulations. Regulatory standards which cover emissions into the outside environment from a workplace and disposal of hazardous wastes from job sites.

epidemiology. The study of occurrence and distribution of disease throughout a population.

equipment room. The last stage or room of the worker decontamination system before entering the work area.

establishing responsibility. An asbestos program manager is designated and is given the responsibility for directing and managing asbestos control program activities.

eyepiece. A component of a full facepiece respirator which is a gas-tight transparent window through which the wearer may see.

facepiece. The portion of a respirator which covers the wearer's nose, mouth, and eyes in a full facepiece.

fallout. The intermittent release of fibers which occurs as a result of weakened bonds in the material, or because of deterioration.

f/cm^3. Fibers per cubic centimeter of air.

FEV. The maximum volume of air that can be forced from an individual's fully inflated lungs in one second (Forced Expiratory Volume/one second).

fiber containment. Enclosing or sealing off an area having airborne asbestos fibers present so that the fibers will not migrate, resulting in contamination of other areas.

fiber control. Minimizing the amount of airborne fiber generation through the application of amended water onto asbestos-containing material, or enclosure (isolation) of the material.

fiber releasability. The potential for generation of airborne fibers from an asbestos-containing source.

fibrosis. A condition of the lungs caused by the inhalation of excessive amounts of fibrous dust, and marked by the presence of scar tissue.

fibrous aerosol monitor (FAM). A portable survey instrument with the capability of providing instantaneous airborne fiber concentration readings.

fireproofing. Spray- or trowel-applied fire-resistant materials.

friable asbestos. Any materials that contain more than 1% asbestos by weight and can be crumbled, pulverized, or reduced to powder by hand pressure.

full facepiece respirator. A respirator which covers the wearer's entire face from the hairline to below the chin.

FVC. Forced Vital Capacity. The measured quantity of air that can be forcibly exhaled from a person's lungs after full inhalation.

glovebag. Plastic bag-type enclosure placed around asbestos-containing pipe lagging so that it may be removed without generating airborne fibers into the atmosphere.

glove-box (bag). Plastic enclosure placed around a specific operation such as valve to contain small areas of materials for asbestos removal.

grade D air. Breathing air which has between 19.5% and 23% oxygen, no more than 5 mg/m^3 of condensed hydrocarbons, no more than 20 ppm of carbon monoxide, no pronounced odor, and a maximum of 1000 ppm carbon dioxide.

ground fault circuit interrupter. A circuit breaker that is sensitive to very low levels of current leakage from a fault in an electrical system.

ground fault interrupter. A device which automatically deenergizes any high voltage system component which has developed a fault in the groundline.

half mask—high efficiency. A respirator which covers one-half of the wearer's face and is equipped with filters capable of screening out 99.97% of all particles larger than 0.3 μm.

heat cramps. Painful spasms of heavily used skeletal muscles, such as those in the hands, arms, legs, and abdomen, sometimes accompanied by dilated pupils and weak pulse. Results from depletion of the salt content of the body.

heat exhaustion. A condition resulting from dehydration and/or salt depletion, or lack of blood circulation, which is usually characterized by fatigue, nausea, headache, giddiness, clammy skin, and a pale appearance.

heat stress. A bodily disorder associated with exposure to excessive heat.

heat stroke. The most severe of the heat stress disorders, resulting from the loss of the body's ability to sweat, and characterized by hot, dry skin, dizziness, nausea, severe headache, confusion, delirium, loss of consciousness, convulsion, and coma.

HEPA. High-efficiency particulate air (air filter).

HEPA-filtered vacuum. A high-efficiency particulate air-filtered vacuum capable of trapping and retaining 99.97% of all particles larger than 0.3 μm.

holding area. The airlock between the shower room and the clean room in a worker decontamination system.

homogeneous. Evenly mixed and similar in appearance and texture throughout.

hose masks. Respirators that supply air from an uncontaminated source through a strong, large-diameter hose to the facepiece and that do not use compressed air or have any pressure regulating devices.

HVAC system. Heating, Ventilation, and Air Conditioning system usually found in large business and industry facilities.

industrial hygienist. A professional qualified by education, training, and experience to recognize, evaluate, and develop controls for occupational health hazards.

joists. The structural building component on which the flooring or roof rests.

local exhaust ventilation. The mechanical removal of air contaminants from a point of operation.

logbook. An official record of all activities which occurred during a removal project.

lung cancer. An uncontrolled growth of abnormal cells in the lungs which normally results in the death of the host.

make-up air. Supplied or recirculated air to offset that which has already been exhausted from an area.

MCEF. Mixed Cellulose Ester Filter, one of several different types of media used to collect asbestos air samples.

mechanical filter respirator. A respiratory protection device which offers protection against airborne particulates including dusts, mists, metal fumes, and smoke.

medical examinations. An evaluation of a person's health status conducted by a medical doctor.

medical history. A record of a person's past health history, including all the hazardous materials they have been exposed to and also any injuries or illnesses which might dictate their future health status.

mesothelioma. A relatively rare form of cancer which develops in the lining of the pleura or peritoneum, with no known cure.

method 7400. NIOSH sampling and analytical method for fibers using phase-contrast microscopy. Replaces method P & CAM–239.

micron (μm). One millionth of a meter.

mil. Prefix meaning one thousandth.

millimeter. One thousandth of a meter.

mineral wool. A commonly used substitute for asbestos.

MSDS. Material Safety Data Sheet.

MSHA. Mine Safety and Health Administration.

negative pressure. An atmosphere created in a work area enclosure such that airborne fibers will tend to be drawn through the filtration system rather than leak out into the surrounding areas. The air pressure inside the work area is less than that outside the work area.

NESHAPS. National Emission Standards for Hazardous Air Pollutants—EPA Regulation 40 CFR subpart M, part 61.

NIOSH. The National Institute for Occupational Safety and Health, which was established by the Occupational Safety and Health Act of 1970.

NIOSH/MSHA. The official approving agencies for respiratory protective equipment, which test and certify respirators.

numerical value. Refers to the types and percentages of asbestos present in a given sample.

oil-less compressor. An air compressor that is not oil-lubricated and which does not allow carbon monoxide to be formed in the breathing air.

open circuit SCBA. A type of self-contained breathing unit which exhausts the exhaled air to the atmosphere instead of recirculating it.

operations and maintenance plan (OMP). Specific procedures and practices developed for the interim control of asbestos-containing material in buildings until it is removed.

"orange booklets." EPA publications issued in March 1979 titled "Asbestos-Containing Materials in School Buildings: A Guidance Document," parts I and II.

OSHA. The Occupational Safety and Health Administration, which was created by the Occupational Safety and Health Act of 1970; serves as the enforcement agency for safety and health in the workplace environment.

oxygen deficient. Any atmosphere containing less than 19.5% oxygen.

P & CAM–239. A NIOSH sampling and analytical method for measuring airborne fibers using phase-contrast microscopy.

particulate contaminants. Minute airborne particles given off in the form of dusts, smokes, fumes, or mists.

PAT samples. Proficiency Analytical Testing of asbestos samples, conducted through NIOSH for laboratories involved with the analysis of asbestos samples.

PEL. Permissible Exposure Limit, as stated by OSHA.

penetrating encapsulant. Liquid material applied to asbestos-containing material to control airborne fiber release by penetrating into the material and binding its components together.

peritoneum. The thin membrane that lines the surface of the abdominal cavity.

personal protective equipment (PPE). Any material or device worn to protect a worker from exposure to, or contact with, any harmful material or force.

personal sample. An air sample taken with the sampling pump directly attached to the worker, with the collecting filter placed in the worker's breathing zone.

personal protection. Notification and instruction of all workers prior to the beginning of a project as to the hazards associated with the job and what they can do to protect themselves from these hazards.

PF. Protection factor as provided by a respirator, which is determined by dividing the airborne fiber concentration outside of the mask by the concentration inside the mask.

phase contrast microscopy (PCM). An optical microscopic technique which is used for the counting of fibers in air samples, but which does not distinguish fiber types.

pipe lagging. The insulation or wrapping around a pipe.

pleura. The thin membrane surrounding the lungs, which lines the internal surface of the chest cavity.

pneumoconiosis. A condition in the lungs which is a result of having inhaled various dusts and particles for a prolonged period of time.

polarized light microscopy (PLM). An optical microscopic technique used to distinguish between different types of asbestos fibers by their shape and unique optical properties.

polyethylene. Plastic sheeting which is often used to seal off an area in which asbestos removal is taking place, for the purpose of preventing contamination of other areas.

posting. Refers to caution or warning signs which should be posted in any area in which asbestos removal is taking place or where airborne fiber levels may present a health hazard.

powered air-purifying respirator (PAPR). Either a full facepiece, helmet, or hooded respirator that has the breathing air powered to the wearer after it has been purified through a filter.

preconstruction conference. A meeting held before any work begins between the contractor and the building owner, at which time the job specifications are discussed and all details of the work agreed upon.

preconstruction physical. Complete medical examination of an employee before the job begins, to determine whether he/she is fit to perform the functions of employment.

pressure demand airline devices. A respiratory protection device which has a regulator and valve design such that there is a continuous flow of air into the facepiece at all times.

prevalent levels. Levels of airborne contaminants occurring under normal conditions.

prevalent samples. Air samples taken under normal conditions (background samples).

protective clothing. Protective, lightweight garments worn by workers performing asbestos abatement to keep gross contamination off the body.

pulmonary. Pertaining to or affecting the lungs or some portion thereof.

pulmonary function. A part of the medical examination. Tests required to determine the health status of a person's lungs.

"purple book." EPA publication of June 1985 titled "Guidance for Controlling Asbestos-Containing Materials in Buildings, 1985 Edition." This document is a revision of the "blue book."

qualitative fit test. A method of testing a respirator's face-to-facepiece seal by covering the inhalation or exhalation valves and either breathing in or out to determine the presence of any leaks.

rales. An abnormal sound heard from the lungs, which does not necessarily indicate any specific disease.

random sample. A sample drawn in such a way that there is no set pattern, which is designed to give a true representation of the entire population or area.

recordkeeping. Detailed documentation of all program activities, decisions, analyses, and any other pertinent information to a project.

resolution. The ability to distinguish between individual objects, as with a microscope.

respirable. Breathable.

respirator program. A written program established by an employer which provides for the safe use of respirators on job sites.

resuspension. The secondary dispersal or reentrainment of settled fibers which have previously been released by impact or fallout.

rip-out. The actual removal of asbestos-containing materials from a building.

risk. The likelihood or probability of developing a disease, or being hurt, as the result of exposure to a contaminant or a condition.

safety glasses. Protective eye equipment.

scanning electron microscopy (SEM). A method of microscopic analysis which utilizes an electron beam directed at the sample and then collects the beams that are reflected to produce an image from which fibers can be identified and counted.

scanning transmission electron microscopy (STEM). A combination of a transmission microscope with scanning and focusing electron coils so that a beam of electrons can be scanned over the sample or pinpointed in a particular area.

SCBA. Self-Contained Breathing Apparatus.

serpentine. One of the two major groups of minerals from which the asbestiform minerals are derived, distinguished by their tubular structure and chemical composition.

shower room. A room between the clean room and the equipment room in a worker decontamination system, in which workers take showers when leaving the work area.

spirometer. An instrument which measures the volume of air being expired from the lungs.

steel beams. Building components which support the joists.

structural member. Any load-supporting member, such as beams and load-supporting walls of a facility.

structural steel. A building component which is designed to support other structural members in a building.

substrate. The material or existing surface located under or behind the asbestos-containing material.

supplied air. A respirator that has a central source respirator of breathing air which is supplied to the wearer by way of an airline.

surfactant. A chemical wetting agent added to water to improve its penetration abilities into asbestos-containing materials.

TLV. Threshold Limit Values: levels of contaminants established by the American Conference of Governmental Industrial Hygienists, to which it is believed that workers can be exposed with minimal adverse health effects.

transmission electron microscopy (TEM). A method of microscopic analysis which utilizes an electron beam that is focused onto a thin sample. As the beam penetrates (transmits) through the sample, the difference in densities produces an image on a fluorescent screen from which samples can be identified and counted.

treated cellulose. An insulation material made of paper or wood products with fire-retarding treatment added.

tumor. A swelling or growth of cells and tissue in the body which does not serve a useful purpose.

TWA. Time-Weighted Average, as in air sampling.

type B reader. A physician with specialized training in reading X-rays, specifically in recognizing lung disorders.

type C supplied-air respirator. A respirator designed to provide a very high level of protection which supplies air to the wearer from an outside source such as a compressor.

USEPA. United States Environmental Protection Agency.

Vermiculite. A micaceous mineral, sometimes used as a substitute for asbestos, that is lightweight and highly water-absorbent.

visible emissions. Airborne fibers given off from an asbestos-containing source that are visible to the human eye.

visual inspection. A walk-through type inspection of the work area to detect incomplete work, damage, or inadequate cleanup of a worksite.

washroom. A room between the work area and the clean room in the equipment decontamination enclosure system where workers shower.

water damage. Deterioration or delamination of ceiling or wall materials due to leaks from plumbing or cracks in the roof.

WBGT. Wet Bulb Globe Temperature, a heat stress index.

wet cleaning. The process of eliminating asbestos contamination from surfaces and objects by using cloths, mops, or other cleaning tools which have been dampened with water.

wetting agents. Materials that are added to water used for wetting the asbestos-containing material in order for the water to penetrate more effectively.

workman's compensation. A system of insurance required in some states by law, financed by employers, which provides payments to employees or their families for occupational injuries, illnesses, or fatalities resulting in loss of wage or income incurred while at work.

Appendix A
Substitutes for Asbestos Materials

Table 1. Substitutes for Asbestos Pipe Insulation

Trade Name		Manufacturer	Substitute Material
1.	Aerotube	Johns-Manville	Foamed Plastic
2.	Alpha-Maritex Style 1925	Alpha Associates	Fibrous Glass
3.	Alpha-Maritex #3111-RW	Alpha Associates	Fibrous Glass
4.	Armaflex 22	Armstrong Cork Co.	Foamed Plastic
5.	CPR	Upjohn	Plastic
6.	Crown	Fibreglass Ltd.	Fibrous Glass
7.	Fit-Rite	Fibrous Glass Products, Inc.	Fibrous Glass
8.	Flame-Safe	Johns-Manville	Fibrous Glass
9.	Glo-Brite	Glo-Brite Products	Poly Foam
10.	GPC	Johns-Manville	Fibrous Glass
11.	Hewflex	H. E. Werner, Inc.	Polyurethane Foam
12.	Kaowool	Babcock & Wilcox	Ceramic Fiber
13.	Kaylo 10	Owens-Corning Fiberglass	Fibrous Glass
14.	Micro-Lok 650	Johns-Manville	Fibrous Glass
15.	Pabco Super Caltemp Type NA	Fibreboard Corp.	Diatomaceous Earth, Nonasbestos Fiber, Lime
16.	PF-CG	Owens-Corning Fiberglass	Fibrous Glass
17.	PMF	Jim Walter Resources, Inc.	Fibrous Glass
18.	Ruberoid Fiber Glass Pipe Insulation	Ruberoid	Fibrous Glass
19.	Snap-On	CertainTeed	Fibrous Glass
20.	TGA-1000	Alpha Associates	Tedlar and Glass Fiber
21.	Therma-K	Ehret Magnesia Mfg. Co.	Glass Fiber

Table 1. Substitutes for Asbestos Pipe Insulation, cont.

Trade Name	Manufacturer	Substitute Material
22. Thermashield	Tecknit	Ceramic Fiber
23. Transifoam	Johns-Manville	Polystyrene
24. Transitop	Johns-Manville	Wood Fiber
25. Uni-Jac	Pittsburgh Corning	Glass Fabric
26. Vapo-Lok	MMM Div. Insular Prods. Corp.	Expanded Polystyrene
27. VB-Vapor Barrier	Johns-Manville	Kraft Paper with Glass Fiber
28. VBL-Vapor Barrier	Johns-Manville	Fibrous Glass
29. VL	Riva and Mariani	Cellular Glass
30. Z-Lock	Fibreglass Ltd.	Fibrous Glass
31. Zonolite	W. R. Grace and Co.	Vermiculite

Source: Courtesy of The National Asbestos Council and M. L. Demyanek at Georgia Institute of Technology.

Table 2. Substitutes for Sprayed-On Asbestos Insulation

Trade Name	Manufacturer	Substitute Material
1. Cafco	USM	Mineral Fibers
2. Cafcote H	USM	Mineral Fibers (also abrasion resistant)
3. Ceramafiber	USM	Ceramic Fiber
4. Ceramospray	Spraycraft Corp.	Ceramic Fiber
5. Ceramwool	Johns-Manville	Ceramic Fiber
6. Encagel V	Childers Products Co.	Urethane
7. Ensolite	U.S. Rubber Co.	Polyvinyl Chloride
8. Ensolite Type M	U.S. Rubber Co.	Polyvinyl Chloride
9. K-13	National Cellulose Corp.	Cellulose

Source: Courtesy of The National Asbestos Council and M. L. Demyanek at Georgia Institute of Technology.

Table 3. Substitutes for Asbestos-Containing Panels or Wallboards

	Trade Name	Manufacturer	Substitute Material
1.	Bestwall	Georgia Pacific	Gypsum
2.	Cal-Shake	U.S. Gypsum	Calcium Silicate
3.	Careytemp 1500	Celotex	Expanded Perlite
4.	Cellofoam	USM	Polystyrene
5.	Cellutron	Owens-Corning	Cellulose
6.	Celot-Therm	Celotex	Perlite
7.	Ceramfab	USM	Ceramic Fiber
8.	Delta-T	Keene Corp.	Ceramic Fiber
9.	Doraspan	Dow	Ceramic Fiber
10.	Dylite	Sinclair-Koppers	Molded Foam
11.	Econacoustic	Sinclair-Koppers	Wood Fiber
12.	Filomat-D	Alpha Associates	Glass Fiber
13.	Fire Shop	Cotton, Inc.	Treated Cotton
14.	Firetard Type X	Johns-Manville	Gypsum
15.	Foamgrid	USM	Polystyrene Foam
16.	Foamsil-28	Pittsburgh Corning	Glass Foam
17.	Foamthane	Pittsburgh Corning	Polyurethane Foam
18.	SE Armalite	Armstrong Cork Co.	Polystyrene
19.	Styrofoam	Dow	EPDM and Aramid
20.	Waterlite Backer	Johns-Manville	Gypsum

Source: Courtesy of The National Asbestos Council and M. L. Demyanck at Georgia Institute of Technology.

Table 4. Substitutes for Asbestos-Containing Fabrics

Trade Name		Manufacturer	Substitute Material
1.	Alpha-Maritex #3111-RW	Alpha Associates	Glass Fiber
2.	Alpha-Maritex #84205	Alpha Associates	Glass Fiber
3.	Aramid	DuPont	Synthetic
4.	Ceel-Tite	Ceel-Co	ABS Plastic
5.	Cerafelt	Johns-Manville	Ceramic Fiber
6.	Ceramfab	USM	Ceramic Fiber
7.	Fiberfrax	The Carborundum Co.	Ceramic Fiber
8.	Fiberseal	Pyrotek, Inc.	Glass Fiber
9.	Fire Stop	Cotton, Inc.	Treated Cotton
10.	Flexfelt	General Insulating	Rock Wool
11.	Flextra	Raybestos-Manhattan	Cotton/Aramid
12.	Fyrepel	Fyrepel	Glass Fiber
13.	Glas Ply	Johns-Manville	Glass Fiber
14.	Glassbestos	Raybestos-Manhattan	Glass Fiber
15.	Glass Web	Steiner Industries	Glass Fiber
16.	GVB Glass/Cloth Vapor Barrier	Johns-Manville	Glass Fiber
17.	Hansoquilt	Baldwin-Ehret-Hill	Glass Fiber
18.	Insulfas	Benjamin Foster	Glass
19.	Kynol	American Kynol Corp.	Novoloid Fiber
20.	Nomex	DuPont	Synthetic
21.	Nor-Fab	Hitco (Armco)	Synthetic
22.	Nor-Fab	AMATEX Corp.	Synthetic
23.	Preox	Gentex Corp.	Heat-Stabilized Polyacrylonitrile
24.	Pyroglas	Raybestos-Manhattan	Glass Fiber
25.	SF 2600	Santa Fe Textiles Inc.	Ceramic Fiber

Table 4. Substitutes for Asbestos-Containing Fabrics, cont.

Trade Name		Manufacturer	Substitute Material
26.	Sisalkraft	St. Regis Paper Co.	Kraft Paper/ Aluminum
27.	Snap Form	CertainTeed	Polyvinyl Chloride
28.	Tempo	Tempo Glove Manufacturing Inc.	Glass Fiber or Leather
29.	Terrybest	A-Best Co.	Kevlar 29
30.	Thermafiber	U.S. Gypsum	Perlite
31.	Thermobest	A-Best Co.	Kevlar and Other Synthetics
32.	Thermo-Ceram	Garlock, Inc.	Ceramic Fiber
33.	Thermoglass	Amatex Corp.	Glass Fibers
34.	Thermo-Sil	Garlock, Inc.	Glass Fibers
35.	Zonolite Dyfoam	W. R. Grace and Co.	Polystyrene

Source: Courtesy of The National Asbestos Council and M. L. Demyanek at Georgia Institute of Technology.

Table 5. Substitutes for Asbestos-Containing Cements/Plasters

	Trade Name	Manufacturer	Substitute Material
1.	Alumino-Hi-Temp	Carey	Alumina
2.	Careytemp 1500	Celotex	Expanded Perlite
3.	Cem-Fil	Asahi Glass Co. Ltd.	Glass Fiber
4.	Cerablanket	Johns-Manville	Ceramic Fiber
5.	Cerachrome	Johns-Manville	Ceramic Fiber
6.	Epitherm 1200	Eagle-Picher	Polyvinyl Chloride
7.	Feldina	Nonco Corp.	Nonasbestos Mineral
8.	Fesco Board	Johns-Manville	Perlite
9.	Mono-Block	Keene Corp.	Mineral Wool
10.	MW-One Insulating Cement	Celotex	Mineral Wool
11.	MW-to	Celotex	Mineral Wool
12.	Nonpariel	Armstrong Cork Co.	Rock Wool
13.	Pabco No. 127	Fibreboard Corp.	Mineral Wool
14.	Pabco Super Caltemp Type NA	Fibreboard Corp.	Diatomaceous Earth, Mineral Fibers, Lime
15.	Super 1900	Keene Corp.	Mineral Wool

Source: Courtesy of The National Asbestos Council and M. L. Demyanek at Georgia Institute of Technology.

Table 6. Substitutes for Asbestos-Containing Brake Lining Discs

Trade Name	Manufacturer	Substitute Material
1. Aramid	DuPont	Synthetic
2. Kynol	American Kynol Corp.	Novoloid Fiber
3. Metal-Might	Lear Siegler, Inc.	Metallic Fiber
4. Premium	Euclid Industries	Synthetic
5. Scan-Pac	Scan-Pac	Metal Chips
6. Star Line	Abex Corp.	Glass Fiber

Source: Courtesy of The National Asbestos Council and M. L. Demyanek at Georgia Institute of Technology.

Table 7. Substitutes for Asbestos-Containing Packing or Fillers

Trade Name		Manufacturer	Substitute Material
1.	CA-5	U.S. Gypsum	Calcium Silicate
2.	Garfite	Garlock, Inc.	Graphite
3.	GFO Fiber	W. L. Gore & Associates, Inc.	PTFE & Graphite
4.	Navalon	Johns-Manville	Ramie
5.	Parfab	Parker Seals	Synthetic
6.	Partherm	Parker Seals	Glass Fiber
7.	Processed Mineral Fiber	Jim Walter Resources, Inc.	Glass Fiber
8.	Snow White	U.S. Gypsum	Calcium Sulfate
9.	Spandreline	PPG	Ceramic Fiber
10.	Spandrelite	PPG	Glass Fiber
11.	Spinsulation	Johns-Manville	Glass Fiber
12.	Style 50-50	Garlock, Inc.	EPDM & Aramid
13.	Synethepak	Garlock, Inc.	Polymer Fiber
14.	Technifoam	Celotex	Urethane

Source: Courtesy of The National Asbestos Council and M. L. Demyanek at Georgia Institute of Technology.

Table 8. Substitutes for Asbestos-Containing Gaskets

Trade Name		Manufacturer	Substitute Material
1.	Aramid	DuPont	Synthetic
2.	Blue-Gard	Garlock, Inc.	Styrene Foam
3.	Chevron	Garlock, Inc.	Fibrous Glass
4.	Fil-Tec	Fil-Tec	Glass Fiber
5.	Garthane	Garlock, Inc.	Graphite
6.	Gylon	Garlock, Inc.	Glass Fiber
7.	Kynol	American Kynol Corp.	Novoloid Fiber
8.	Marblock	Garlock, Inc.	Glass Wool
9.	Nobestos	Rogers Corp.	Chloroprene, Nitrile, Acrylic
10.	Prolene	Garlock, Inc.	EPDM Rubber
11.	Temp Mat	Pittsburgh Corning	Glass Fiber
12.	Texo	PPG	Fiber Glass

Source: Courtesy of The National Asbestos Council and M. L. Demyanek at Georgia Institute of Technology.

Appendix B

Asbestos Information Sources

Information Sources

A/C Pipe Producers Association
1600 Wilson Blvd.
Suite 1308
Arlington, VA 22209
(703) 841-1556

American Industrial Health Council
1330 Connecticut Ave., NW
Washington, DC 20036
(202) 659-0060

American Industrial Hygiene Association
475 Wolf Ledges Parkway
Akron, OH 44311
(216) 762-7294

American Lung Association
1740 Broadway
New York, NY 10019-4374
(212) 315-8700

APCA (formerly Air Pollution Control Association)
PO Box 2861
Pittsburgh, PA 15230
(412) 232-3444

Asbestos Abatement Council
25 K Street, NE

Washington, DC 20002
(202) 783-2924

Asbestos Action Program
TS 794
US Environmental Protection Agency
401 M Street, SW
Washington, DC 20460
(202) 382-3949

Asbestos Council of the Midwest
199 Pierce Street, Suite 204
Birmingham, MI 48011
(313) 642-9797

Asbestos Information Association
1745 Jefferson Davis Highway
Arlington, VA 22202
(703) 979-1150

Asbestos Information Center
Tufts University
Curtis Hall
474 Boston Avenue
Medford, MA 02155
(617) 381-3531

Asbestos Information Centre Ltd.
22-28 High Street
Epsom, Surrey KT19 8AH, England
037-274-2055

The Asbestos Institute
1130 Sherbrooke Street West
Suite 410
Montreal, Quebec, Canada, H3A 2M8
(514) 844-3956

Asbestos International Association
68 Gloucester Place
London W1H 3H1, England
01-486-3528/9

Asbestos Litigation Group
174 East Bay Street, Suite 100
Charleston, SC 29401
(803) 577-6747

Asbestos Removal Contractors Association
45 Sheen Lane
London SW148AB, England
01-876-4415

Asbestos Victims of America
PO Box 559
Capitola, CA 95010
(408) 476-3646

Association of School Business Officials–International
11401 North Shore Drive
Reston, VA 22090-4232

ASTM Building Construction Committee
1916 Race Street
Philadelphia, PA 19103
(215) 299-5496

Building Owners and Managers Association International
1250 Eye Street, Suite 200
Washington DC 20005
(202) 289-7000

Government Refuse Collection and Disposal Association
PO Box 7219
Silver Spring, MD 20910
(301) 585-2898

International Association of Heat and Frost
Insulators and Asbestos Workers
1300 Connecticut Ave., NW, Suite 505
Washington, DC 20036
(202) 785-2388

Mid-Atlantic Asbestos Training Center
Rutgers Medical School
675 Hoes Lane
Piscataway, NJ 98854
(201) 463-4500

Midwest Asbestos Information Center
University of Illinois at Chicago
Box 6998
Chicago, IL 60680
(312) 996-5762

National Asbestos Council
2786 North Decatur Road, Suite 220
Decatur, GA 30033
(404) 292-3802

National Asbestos Training Center
University of Kansas
Division of Continuing Education
5005 West 95th Street
Shawnee Mission, KS 66207
(913) 648-5790

National Association of Demolition Contractors
4415 West Harrison Street
Hillside, IL 60162
(312) 449-5959

National Building Material Distributors Association
1417 Lake Cook Road, Suite 130

Deerfield, IL 60015
(312) 945-6940

National Insulation Contractors Association
1025 Vermont Ave., NW, Suite 410
Washington, DC 20005
(202) 783-6277

National Organization for Improving School Environments
12043 Greywing Square, Apt. C-3
Reston, VA 22091
Telephone not available

National Solid Wastes Management Association
1730 Rhode Island Ave., NW, Suite 1000
Washington, DC 20036
(202) 463-4613

National Institute for Occupational Safety and Health
Respiratory Disease Study Division
944 Chestnut Ridge Road
Morgantown, WV 26505
(304) 291-4474

Professional Association for Asbestos Control
7709 West Beloit Road
Milwaukee, WI 53219
(414) 541-7744

Pacific Asbestos Information Center
2223 Fulton Street
Berkeley, CA 94720
(415) 643-7143

Rocky Mountain Center for Occupational and
Environmental Health
University of Utah, Building 512
Salt Lake City, UT 84112
(801) 581-5710

Safe Buildings Alliance
655 Fifteenth St., NW, Suite 1200
Washington, DC 20005
(202) 879-5120

Safety Equipment Distributor's Association
111 East Wacker Drive, Suite 600
Chicago, IL 60601
(312) 644-6610

Society for the Prevention of Asbestosis
and Industrial Diseases
38 Drapers Road
Enfield, Middlesex EN2 8UL, England
01-366-1640

Solid Waste Processing Division
American Society of Mechanical Engineers
345 E. 47th Street
New York, NY 10017
(212) 705-7159

Southeastern Asbestos Information Center
Georgia Institute of Technology
Atlanta, GA 30332
(404) 894-3806

US Environmental Protection Agency
Office of Toxic Substances
401 M Street, SW
Washington, DC 20460
(202) 382-3813

EPA Region Offices

Region 1:
USEPA
JFK Federal Building

Boston, MA 02203
(617) 223-0585
NESHAPS: (617) 223-4872

Region 2:
USEPA
Woodbridge Avenue
Edison, NJ 08837
(201) 321-6668
NESHAPS: (212) 264-4479

Region 3:
USEPA
841 Chestnut Street
Philadelphia, PA 19107
(215) 597-9859
NESHAPS: (215) 597-6552

Region 4:
USEPA
345 Courtland Street NE
Atlanta, GA 30365
(404) 881-3864
NESHAPS: (404) 881-4901

Region 5:
USEPA
230 S. Dearborn Street
Chicago, IL 60604
(312) 886-6879
NESHAPS: (312) 353-2088

Region 6:
USEPA
First International Building
1201 Elm Street
Dallas, TX 75270
(214) 767-5314
NESHAPS: (214) 767-9835

Region 7:
USEPA
726 Minnesota Avenue
Kansas City, KS 66101
(913) 236-2838
NESHAPS: (913) 236-2576

Region 8:
USEPA
999 18th Street
Denver, CO 80202
(303) 293-1730
NESHAPS: (303) 293-1767

Region 9:
USEPA
215 Fremont Street
San Francisco, CA 94105
(415) 974-8588
NESHAPS: (415) 974-7648

Region 10:
USEPA
1200 Sixth Avenue
Seattle, WA 98101
(206) 442-2632
NESHAPS: (206) 442-2724

Appendix C
AHERA Training Requirements

The 1986 Asbestos Hazard Emergency Response Act (AHERA) directed the Environmental Protection Agency (EPA) to develop rules for the assessment and management of asbestos in our nation's schools. The Model Accreditation Plan, a final rule, was published concurrently with the proposed rules on Asbestos Containing Materials (ACM) on April 30, 1987, 40 CFR Part 763 (Appendix C). The Model Accreditation Plan specifies training and accreditation requirements for five categories of personnel involved in asbestos hazard control in schools.[25] These five disciplines are: inspectors, management planners, abatement project designers, asbestos abatement contractors and supervisors, and asbestos abatement workers. These disciplines are listed in the order that hazard management progresses: inspections first, followed by management plan development, project design, and finally, actual abatement. In addition to accreditation for the five categories in the Model Plan, the Proposed Rules contain provisions for training of operations and maintenance workers. The Model Accreditation Plan is divided into four units. Unit 1 details the training requirements for each of the categories listed above. Unit 2 addresses EPA approval of individual state contractor accreditation programs. Unit 3 discusses EPA approval of training courses. Unit 4 explains the interim accreditation process.

TRAINING REQUIREMENTS

Inspectors

Inspectors are those professionals who identify ACM and assess its condition. The Model Plan requires a three-day training course with at least four hours of hands-on training for inspectors. Individual respirator fit testing and a written 50-question multiple-choice exam must be included. The following topics are to be addressed:

- background information on asbestos
- health effects related to asbestos exposure
- role of the inspector
- legal issues
- building systems
- public/employee/building occupant relations
- preinspection planning
- inspection and assessment procedures
- bulk sampling/documentation
- respiratory and other protective equipment
- recordkeeping and report writing
- regulations
- field trip
- course review

Management Planners

Management planners use the data reported by inspectors to assess the ACM's hazard, determine appropriate response actions, and develop a schedule for implementing those actions. Management planners must take the inspector course plus a two-day management planning training course. They must pass a 50-question multiple-choice exam (in addition to the inspector exam).

The following topics will be included in the course:

- course overview
- evaluation/interpretation of survey results
- hazard assessment
- legal implications

- evaluation and selection of control options
- role of other professionals
- operations and maintenance plans
- regulations
- recordkeeping
- assembling the management plan
- course review

Abatement Project Designers

This group is responsible for the design of the abatement project. They determine exactly how the abatement work should be conducted. States have the option of requiring either a three-day abatement project designer course, or the four-day designer course which culminates with a 100-question multiple-choice examination.

Topics that are covered must include:

- background information on asbestos
- potential health effects
- overview of abatement construction projects
- safety system design specifications
- field trip
- employee personal protective equipment
- safety hazards other than asbestos
- fiber aerodynamics and control
- designing abatement solutions
- budgeting/cost estimating
- writing abatement specifications
- preparing abatement drawings
- contract preparation and administration
- legal and insurance considerations
- replacement of asbestos
- role of other professionals
- design of projects in occupied buildings
- regulations
- course review

Asbestos Abatement Contractors and Supervisors

Contractors and supervisors are responsible for carrying out the abatement work. This group includes the immediate supervisor of the abatement workers. The Model Plan requires that the program include at least six hours of hands-on training, individual respirator fit testing, and a 100-question multiple-choice exam.

Subjects to be covered are:

- physical characteristics of asbestos and ACM
- potential health effects related to asbestos exposure
- employee personal protective equipment
- state-of-the-art work practices
- personal hygiene and decontamination
- safety hazards other than asbestos
- medical surveillance
- air monitoring
- regulations
- respirator and medical surveillance programs
- insurance and liability issues
- recordkeeping
- supervisory techniques to encourage safe work practices
- contract specifications
- course review

Asbestos Abatement Workers

Asbestos abatement workers physically carry out the abatement effort. Workers must take a three-day training course with a 50-question multiple-choice exam. At least six hours of hands-on training, including individual respirator fit testing, is required.

The following topics are to be included in the asbestos abatement worker course:

- physical characteristics of asbestos
- potential health effects
- employee personal protective equipment
- state-of-the-art work practices
- personal hygiene and decontamination
- safety hazards other than asbestos

- medical monitoring
- air monitoring
- regulation
- respiratory protection programs
- course review

Operations and Maintenance

Accreditation of operations and maintenance workers is not required. The Proposed Rules will require maintenance and custodial staff who work in a building containing ACM to take a two-hour awareness training course.

Awareness training includes:

- background information on asbestos
- health effects
- locations of asbestos in each building
- recognition of damage to ACM
- name and availability of responsible person
- availability and location of management plan

Maintenance and custodial workers who conduct activities that will result in the disturbance of ACM will be required to receive an additional 14 hours of training including the following topics:

- proper methods of handling ACM
- respiratory protection
- personal protective equipment
- regulations
- hands-on training

APPROVAL OF STATE ACCREDITATION PROGRAMS AND TRAINING COURSES

As discussed in Unit 11 of the Model Plan, each state may adopt an accreditation program of its own. To receive EPA approval, a state's programs must be at least as stringent as the Model Plan. States may require additional credentials if they wish.

Some training courses currently have "EPA approval." Up to this time, EPA has approved courses using criteria different from those in the Model Plan. Some of these courses may qualify for approval under AHERA, or portions of the program may need to be modified.

The current "EPA-approved" courses will likely qualify for interim accreditation under the provision of Unit 1V. This will allow persons who passed those courses after January 1, 1985 to work for one year after their state establishes its accreditation programs. (Most states will adopt their programs by summer 1988).

Once a contractor's interim accreditation has expired, he or she must become fully accredited. To achieve full accreditation the contractor must pass the appropriate course(s) and fulfill any additional state requirements. In addition, many states have requirements for contract certification that extends beyond schools. There are also municipal laws governing asbestos in many areas. For example, New York City passed a law in April 1987 affecting the removal of asbestos in residential and commercial buildings. Similar legislation is pending in several other major cities.

Appendix D

Typical Equipment Costs Based on 1987 Catalog Prices[22]

Table 1.

Item	Quantity	Cost &	Unit
Coverall, zipper front	25	54.00	case
Coverall, zipper front, attached hood	25	62.50	case
Coverall, zip front, attached hood & boot, elas wrist	25	78.10	case
Coverall, plain Tyvek, hood & bts., mylar face shld, air hose	25	5.08	each
Hood, plain Tvyek	250	113.62	case
Profo-Briefs, plain Tyvek	100	53.35	case
Boot covers, plain Tyvek	250	115.50	case
Gloves, Latex, unlined, large	12 doz per case	7.60	doz
Gloves, Latex, unlined, X-large	12 doz per case	7.95	doz
Gloves, cotton knit wrist	12 doz per case	8.40	doz
Gloves, PVC-coated, knit-wrist	12 doz per case	14.50	doz
Gloves, leather palm	12 doz per case	24.00	doz

Table 1, cont.

Boots & Safety Gear			
Boot covers, Latex	50	4.68	each
Boot covers, vinyl, nonskid soles	1	5.25	pair
Boots, plain toe, PVC knee boot	1	11.00	pair
Boots, steel toe, PVC knee boot	1	13.00	pair
Boots, slush, yellow	1	10.50	pair
Hard hat, AO	1	4.50	each
Eyeglasses, safety AO	1	6.35	each
Poly			
Polyethylene film, fire retardant, 6-mil, 20 × 100	1	69.95	roll
Polyethylene film, fire retardant, reinforced, 6-mil 20 × 100	1	175.00	roll
Glue Gun, Tape & Adhesives			
Duct tape, 2 in. × 60 yd	24 rolls case	86.64	case
Spray adhesive, High tack	12 rolls case	67.80	case
Spray adhesive, Regular	12 rolls case	55.68	case
Waste Bags			
33 × 50, 6-mil yellow, "Danger"	75 cs	45.00	
33 × 40, 6-mil yellow, "Danger"	100 cs	40.50	

Table 1, cont.

Waste Bags, cont.			
44 × 46, clear, "Danger"	65 cs	42.25	
38 × 63, 6-mil yellow, "Danger"	50 cs	0.00	
33 × 50, clear, no printing	75 cs	45.00	
Label, "Asbestos Danger," 3 × 5, 500/roll	1	50.00	roll
Sign, paper, "Asbestos Danger," 11 × 17, 200/pack	1	33.00	pack
Sign, fiberglass, "Asbestos Danger," 14 × 20	1	13.55	each
Label, "Asbestos Free," 500/roll	1	31.20	roll
Tape, "Danger, Asbestos Removal"	1	44.84	roll
Sign, "Respirators Required In Area"	1	13.55	each
Label, "Asbestos Danger," for maint. workers, 3 × 5 roll	1	50.00	roll
Tools			
Utility knife, disposable	1	1.45	
Utility knife, adjustable	1	3.81	
Scraper, 3-in. bent blade	1	5.83	
Scraper, 3-in. chisel blade	1	5.43	
Light stand, portable	1	259.47	
Surfactant concentrate, 5 gal	1	130.08	each
Misting spray bottle, 1 qt	1	1.79	each
Spray bottle, 3.0 gal	1	32.20	each
Encapsulant, 5 gal	1	97.00	each
Encapsulant, 1 qt	1	19.95	each

Table 1, cont.

Tools, cont.			
Encapsulant, high-temp, fire retard, 5 gal	1	99.00	each
Neg Air Equipment			
30611 Filter, main, HEPA 24 × 24 × 11.5	1	159.75	each
30615 Filter, #2 prefilter	6 per case	57.00	case
30616 Filter, #3 prefilter	30 per case	29.90	case
30607 Flex duct, 12 in. × 25 feet	1	85.00	each
30748 Hydro-Safe Ground-Fault Interrupt	1	75.00	each
Sample Filters			
Millipore, 37 mm	50 per box	90.60	box
Millipore, 25 mm, cowl	50 per box	120.00	box
Nuclepore, 37 mm	25 per box	48.00	box
Single chamber water filtration system	1	238.90	
Filter bag, 5 micron	1	4.42	
Filter bag, 10 micron	1	4.42	
Pressure-Demand Face Pieces for Type-C Systems			
Full-face, pressure demand face piece		270.00	each
New Egress Cartridge Face Pieces for Type-C Systems			
Full-face supplied air mask, egress cartridges		196.20	each

Table 1, cont.

Supplies & Accessories for Supplied Air Systems		
Egress HEPA cartridge, 6 per pack	55.20	pack
Extra manifold panels for high-pressure system	590.00	each
Carbon monoxide tubes	1.80	each
Toxic gas detection kit	185.00	each
Type C Air Purifying System		8,900.00

Bibliography

1. G. Parkinson, "Asbestos Substitutes," *Chem. Eng.* 18–20 (October 27, 1986).
2. "News Front Feature," *Chem. Eng.* 23 (October 27, 1986).
3. *Federal Register,* 51(119) (June 20, 1986).
4. S.L. Biegel, "Asbestos Abatement by Design," NAC4(3)30033 (Decatur, GA: National Asbestos Council, 1986).
5. Author's unpublished lecture notes.
6. Lists furnished by Eva Clay, 1987 Chairperson, National Asbestos Council, with contributions by M.L. Demyanek of the Georgia Institute of Technology.
7. National Fire Protection Association—Standard 220 and 241 (Quincy, MA: National Fire Protection Association, 1986).
8. EPA 560/5-85-024, 3:1–6 (Washington, DC: U.S. Environmental Protection Agency).
9. "Guidance for Controlling Asbestos-Containing Materials in Buildings," Appendix K (Washington, DC: U.S. Environmental Protection Agency, 1985).
10. "Model EPA Curriculum," prepared by Environmental Sciences, Inc., and Georgia Tech Research Institute, draft release (1988).

11. "Guidance for Controlling Asbestos-Containing Materials in Buildings," Appendix H (Washington, DC: U.S. Environmental Protection Agency, 1985).
12. "Asbestos Abatement Training Program," Grant CX813167-01-0 Training Manual (1986) G:131–140 (Washington, DC: U.S. Environmental Protection Agency).
13. Based on U.S. Navy Specifications Number 02075N, Washington, DC (1987).
14. *Federal Register*, 51(119) (June 20, 1986).
15. OSHA Section 1910.1001:911 (Washington, DC: Occupational Safety and Health Administration, 1986).
16. OSHA Section 1926.58: Appendix I (Washington, DC: Occupational Safety and Health Administration, early 1986 release).
17. NIOSH Publication Number 85-115, sections 2–3, 8 (National Institute of Occupational Safety and Health).
18. State of Ohio, Department of Health, Asbestos Regulations Draft, Columbus, OH (March, 1986).
19. *Federal Register*, 51(119) (June 20, 1986).
20. *Federal Register*, 51(119) (June 20, 1986).
21. Author's unpublished lecture notes.
22. Costs based on phone calls to advertisers (summer, 1986); *National Asbestos Council Journal*; and ACTI catalog (1986).
23. "Smith's Law Review Series," Torts 1975, Contracts 1976 (Hill, Rossen & Sogg).
24. Listing provided by Tom Laubenthal, Training Director, National Asbestos Council, July, 1988.
25. K. Hays and S. Millman, *National Asbestos Council Journal*, 5(3):28–31 (Summer 1987).

Index